T0282307

LONDON MATHEMATICAL SOCIETY LECTURE NOTE SERIES

The books in the series listed below are available from booksellers, or, in case of difficulty, from Cambridge University Press.

34 Representation theory of Lie groups, M.F. ATIYAH *et al*
36 Homological group theory, C.T.C. WALL (ed)
39 Affine sets and affine groups, D.G. NORTHCOTT
46 p-adic analysis: a short course on recent work, N. KOBLITZ
49 Finite geometries and designs, P. CAMERON, J.W.P. HIRSCHFELD & D.R. HUGHES (eds)
50 Commutator calculus and groups of homotopy classes, H.J. BAUES
57 Techniques of geometric topology, R.A. FENN
59 Applicable differential geometry, M. CRAMPIN & F.A.E. PIRANI
66 Several complex variables and complex manifolds II, M.J. FIELD
69 Representation theory, I.M. GELFAND *et al*
74 Symmetric designs: an algebraic approach, E.S. LANDER
76 Spectral theory of linear differential operators and comparison algebras, H.O. CORDES
77 Isolated singular points on complete intersections, E.J.N. LOOIJENGA
79 Probability, statistics and analysis, J.F.C. KINGMAN & G.E.H. REUTER (eds)
80 Introduction to the representation theory of compact and locally compact groups, A. ROBERT
81 Skew fields, P.K. DRAXL
82 Surveys in combinatorics, E.K. LLOYD (ed)
83 Homogeneous structures on Riemannian manifolds, F. TRICERRI & L. VANHECKE
86 Topological topics, I.M. JAMES (ed)
87 Surveys in set theory, A.R.D. MATHIAS (ed)
88 FPF ring theory, C. FAITH & S. PAGE
89 An F-space sampler, N.J. KALTON, N.T. PECK & J.W. ROBERTS
90 Polytopes and symmetry, S.A. ROBERTSON
91 Classgroups of group rings, M.J. TAYLOR
92 Representation of rings over skew fields, A.H. SCHOFIELD
93 Aspects of topology, I.M. JAMES & E.H. KRONHEIMER (eds)
94 Representations of general linear groups, G.D. JAMES
95 Low-dimensional topology 1982, R.A. FENN (ed)
96 Diophantine equations over function fields, R.C. MASON
97 Varieties of constructive mathematics, D.S. BRIDGES & F. RICHMAN
98 Localization in Noetherian rings, A.V. JATEGAONKAR
99 Methods of differential geometry in algebraic topology, M. KAROUBI & C. LERUSTE
100 Stopping time techniques for analysts and probabilists, L. EGGHE
101 Groups and geometry, ROGER C. LYNDON
103 Surveys in combinatorics 1985, I. ANDERSON (ed)
104 Elliptic structures on 3-manifolds, C.B. THOMAS
105 A local spectral theory for closed operators, I. ERDELYI & WANG SHENGWANG
106 Syzygies, E.G. EVANS & P. GRIFFITH
107 Compactification of Siegel moduli schemes, C-L. CHAI
108 Some topics in graph theory, H.P. YAP
109 Diophantine analysis, J. LOXTON & A. VAN DER POORTEN (eds)
110 An introduction to surreal numbers, H. GONSHOR
111 Analytical and geometric aspects of hyperbolic space, D.B.A. EPSTEIN (ed)
113 Lectures on the asymptotic theory of ideals, D. REES
114 Lectures on Bochner-Riesz means, K.M. DAVIS & Y-C. CHANG
115 An introduction to independence for analysts, H.G. DALES & W.H. WOODIN
116 Representations of algebras, P.J. WEBB (ed)
117 Homotopy theory, E. REES & J.D.S. JONES (eds)
118 Skew linear groups, M. SHIRVANI & B. WEHRFRITZ
119 Triangulated categories in the representation theory of finite-dimensional algebras, D. HAPPEL
121 Proceedings of *Groups - St Andrews 1985*, E. ROBERTSON & C. CAMPBELL (eds)
122 Non-classical continuum mechanics, R.J. KNOPS & A.A. LACEY (eds)
124 Lie groupoids and Lie algebroids in differential geometry, K. MACKENZIE
125 Commutator theory for congruence modular varieties, R. FREESE & R. MCKENZIE
126 Van der Corput's method of exponential sums, S.W. GRAHAM & G. KOLESNIK
127 New directions in dynamical systems, T.J. BEDFORD & J.W. SWIFT (eds)

128 Descriptive set theory and the structure of sets of uniqueness, A.S. KECHRIS & A. LOUVEAU
129 The subgroup structure of the finite classical groups, P.B. KLEIDMAN & M.W.LIEBECK
130 Model theory and modules, M. PREST
131 Algebraic, extremal & metric combinatorics, M-M. DEZA, P. FRANKL & I.G. ROSENBERG (eds)
132 Whitehead groups of finite groups, ROBERT OLIVER
133 Linear algebraic monoids, MOHAN S. PUTCHA
134 Number theory and dynamical systems, M. DODSON & J. VICKERS (eds)
135 Operator algebras and applications, 1, D. EVANS & M. TAKESAKI (eds)
136 Operator algebras and applications, 2, D. EVANS & M. TAKESAKI (eds)
137 Analysis at Urbana, I, E. BERKSON, T. PECK, & J. UHL (eds)
138 Analysis at Urbana, II, E. BERKSON, T. PECK, & J. UHL (eds)
139 Advances in homotopy theory, S. SALAMON, B. STEER & W. SUTHERLAND (eds)
140 Geometric aspects of Banach spaces, E.M. PEINADOR and A. RODES (eds)
141 Surveys in combinatorics 1989, J. SIEMONS (ed)
142 The geometry of jet bundles, D.J. SAUNDERS
143 The ergodic theory of discrete groups, PETER J. NICHOLLS
144 Introduction to uniform spaces, I.M. JAMES
145 Homological questions in local algebra, JAN R. STROOKER
146 Cohen-Macaulay modules over Cohen-Macaulay rings, Y. YOSHINO
147 Continuous and discrete modules, S.H. MOHAMED & B.J. MÜLLER
148 Helices and vector bundles, A.N. RUDAKOV et al
149 Solitons, nonlinear evolution equations and inverse scattering, M.J. ABLOWITZ & P.A. CLARKSON
150 Geometry of low-dimensional manifolds 1, S. DONALDSON & C.B. THOMAS (eds)
151 Geometry of low-dimensional manifolds 2, S. DONALDSON & C.B. THOMAS (eds)
152 Oligomorphic permutation groups, P. CAMERON
153 L-functions and arithmetic, J. COATES & M.J. TAYLOR (eds)
154 Number theory and cryptography, J. LOXTON (ed)
155 Classification theories of polarized varieties, TAKAO FUJITA
156 Twistors in mathematics and physics, T.N. BAILEY & R.J. BASTON (eds)
157 Analytic pro-*p* groups, J.D. DIXON, M.P.F. DU SAUTOY, A. MANN & D. SEGAL
158 Geometry of Banach spaces, P.F.X. MÜLLER & W. SCHACHERMAYER (eds)
159 Groups St Andrews 1989 volume 1, C.M. CAMPBELL & E.F. ROBERTSON (eds)
160 Groups St Andrews 1989 volume 2, C.M. CAMPBELL & E.F. ROBERTSON (eds)
161 Lectures on block theory, BURKHARD KÜLSHAMMER
162 Harmonic analysis and representation theory for groups acting on homogeneous trees, A. FIGA-TALAMANCA & C. NEBBIA
163 Topics in varieties of group representations, S.M. VOVSI
164 Quasi-symmetric designs, M.S. SHRIKANDE & S.S. SANE
165 Groups, combinatorics & geometry, M.W. LIEBECK & J. SAXL (eds)
166 Surveys in combinatorics, 1991, A.D. KEEDWELL (ed)
167 Stochastic analysis, M.T. BARLOW & N.H. BINGHAM (eds)
168 Representations of algebras, H. TACHIKAWA & S. BRENNER (eds)
169 Boolean function complexity, M.S. PATERSON (ed)
170 Manifolds with singularities and the Adams-Novikov spectral sequence, B. BOTVINNIK
172 Algebraic varieties, GEORGE R. KEMPF
173 Discrete groups and geometry, W.J. HARVEY & C. MACLACHLAN (eds)
174 Lectures on mechanics, J.E. MARSDEN
175 Adams memorial symposium on algebraic topology 1, N. RAY & G. WALKER (eds)
176 Adams memorial symposium on algebraic topology 2, N. RAY & G. WALKER (eds)
177 Applications of categories in computer science, M.P. FOURMAN, P.T. JOHNSTONE, & A.M. PITTS (eds)
178 Lower K- and L-theory, A. RANICKI
179 Complex projective geometry, G. ELLINGSRUD, C. PESKINE, G. SACCHIERO & S.A. STRØMME (eds)
180 Lectures on ergodic theory and Pesin theory on compact manifolds, M. POLLICOTT
181 Geometric group theory I, G.A. NIBLO & M.A. ROLLER (eds)
182 Geometric group theory II, G.A. NIBLO & M.A. ROLLER (eds)
183 Shintani zeta functions, A. YUKIE
184 Arithmetical functions, W. SCHWARZ & J. SPILKER
185 Representations of solvable groups, O. MANZ & T.R. WOLF
186 Complexity: knots, colourings and counting, D.J.A. WELSH
187 Surveys in combinatorics, 1993, K. WALKER (ed)
190 Polynomial invariants of finite groups, D.J. BENSON

London Mathematical Society Lecture Note Series. 190

Polynomial Invariants of Finite Groups

D. J. Benson
University of Oxford

CAMBRIDGE UNIVERSITY PRESS
Cambridge, New York, Melbourne, Madrid, Cape Town, Singapore,
São Paulo, Delhi, Dubai, Tokyo, Mexico City

Cambridge University Press
The Edinburgh Building, Cambridge CB2 8RU, UK

Published in the United States of America by
Cambridge University Press, New York

www.cambridge.org
Information on this title: www.cambridge.org/9780521458863

First published 1993

A catalogue record for this publication is available from the British Library

Library of Congress cataloguing in publication data available

ISBN 978-0-521-45886-3 Paperback

Contents

Introduction

Invariant theory is a subject with a long history, a subject which in the words of Dieudonné and Carrell [32] "has already been pronounced dead several times, and like the Phoenix it has been again and again arising from the ashes."

The starting point is a linear representation of a linear algebraic group on a vector space, which then induces an action on the ring of polynomial functions on the vector space, and one looks at the ring consisting of those polynomials which are invariant under the group action. The reason for restricting ones attention to linear algebraic group representations, or equivalently to Zariski closed subgroups of the general linear group on the vector space, is that the polynomial invariants of an arbitrary subgroup of the general linear group are the same as the invariants of its Zariski closure.

In the nineteenth century, attention focused on proving finite generation of the algebra of invariants by finding generators for the invariants in a number of concrete examples. One of the high points was the proof of finite generation for the invariants of $SL_2(\mathbb{C})$ acting on a symmetric power of the natural representation, by Gordan [39] (1868). The subject generated a language all its own, partly because of the influence of Sylvester, who was fond of inventing words to describe rather specialized concepts.

In the late nineteenth and early twentieth century, the work of David Hilbert and Emmy Noether in Göttingen clarified the subject considerably with the introduction of abstract algebraic machinery for addressing questions like finite generation, syzygies, and so on. This was the beginning of modern commutative algebra. Hilbert proved finite generation for the case of $GL_n(\mathbb{C})$ acting on a symmetric power of the natural representation [45] (1890), using the fact that a polynomial ring is Noetherian (Hilbert's basis theorem).

Hilbert's fourteenth problem, posed at the 1900 International Congress of Mathematicians in Paris, asks whether the invariants are always finitely generated. Well, actually, this is false. He was under the impression that L. Maurer had just proved this, and he goes on to ask a more complicated question which generalizes this. In fact, Maurer's proof contained a mistake, and Nagata [65] (1959) found a counterexample to finite generation. An account of the Hilbert problems and the progress made on them up to the mid nineteen seventies is contained in the A.M.S. publication [6].

For reductive algebraic groups, however, finite generation does hold. This was first proved over the complex numbers by Weyl [110] (1926), using his "unitarian trick", which amounts to proving that reductive groups are linearly reductive. In characteristic p, reductive groups are no longer necessariy linearly reductive (the only connected linearly reductive groups in characteristic p are the tori), but Mumford [64] conjectured and Haboush [41] (1975) proved that reductive groups are "geometrically reductive", and Nagata [67] had already proved that this is enough to ensure finite generation for the invariants.

The question of when the ring of invariants of representation of a finite group over the complex numbers is a polynomial ring was solved by Shephard and Todd [95] (1954). Their theorem states that the ring of invariants is a polynomial ring if and only if the group is generated by elements fixing a hyperplane. Such elements are called pseudoreflections. Their proof involved classifying the finite groups generated by pseudoreflections, and is a tour de force of combinatorics and geometry. The classification was used to prove that groups generated by pseudoreflections have invariant rings that are polynomial rings, and this first implication together with Molien's theorem was then used in the proof of the converse. Later, Chevalley [26] (1955) provided a proof of the first implication for a real reflection group, which avoided classifying the groups. This proof involved the combinatorics of differential operators. Serre later showed how this proof could be adapted to the complex case. A more homological proof of this first implication was found by Larry Smith [96] (1985), and a proof of the converse which works in arbitrary characteristic appeared in Bourbaki [17] Chapter 5 §5, Exercise 7 (1981). The first implication is false in arbitrary characteristic.

The next interesting ring theoretic property of the ring of invariants of a reductive group is that it is Cohen–Macaulay. This was proved in the case of a finite group by Hochster and Eagon [47] (1971). Watanabe [107, 108] (1974) proved that in this case the ring of invariants is Gorenstein provided the group acts as matrices with determinant one. In the general reductive case, both these statements were proved by Hochster and Roberts [48] (1974).

The question of when unique factorization holds in rings of invariants was first attacked by Samuel [86] (1964). He developed the framework for defining the ideal class group of a Krull domain. A finitely generated Krull domain is the same thing as a normal domain (i.e., a Noetherian integrally closed domain). Using Samuel's ideas, Nakajima proved that the ring of invariants of a finite group is a unique factorization domain if and only if there are no non-trivial homomorphisms from the group to the multiplicative group of the field, taking every pseudoreflection to the identity element.

A lot of the ideas involved in invariant theory in the reductive case are already present in the finite case; and the latter has the advantage that one does not have to spend an inordinate amount of time setting up the machinery of algebraic groups before getting to the interesting theorems. For this reason, I decided to restrict the scope of this book to the case of finite group representations.

This book is based on a lecture course I gave at Oxford in the spring of 1991. My starting point for these lectures was the excellent survey article of Richard Stanley [102], which I strongly recommend to anyone wishing to get an overview of the subject. The influence of this article will be apparent in almost every part of this book. The main direction in which this book differs from that article, apart from length and amount of detail, is that I have tried to indicate how much is true in arbitrary characteristic, whereas Stanley restricts his discussion to characteristic zero. Also, Stanley does not discuss divisor classes and unique factorization.

At Larry Smith's invitation, I spent two months in Göttingen in the summer of 1991, and he persuaded me that we should write a book together on invariant theory of finite groups. Differences in personal style prevented this from being realized as a joint project, but I would like to take this opportunity to thank him for discussions which had a considerable effect on the presentation of the material in Sections 7.2 and 8.1, and indeed for persuading me to embark on this project in the first place.

I would also like to thank Bill Crawley–Boevey for taking the trouble to make many constructive criticisms, and for suggesting the inclusion of the appendices; and David Tranah of Cambridge University Press for his usual endless supply of patience.

Chapter 1

Finite Generation of Invariants

1.1 The basic object of study

Let G be a finite group, K a field of coefficients and V a finite dimensional K-vector space on which G acts by linear substitutions (or equivalently a finitely generated KG-module). We write $K[V]$ for the ring of polynomial functions on V. In other words, if V has dimension n as a vector space, and $x_1, \ldots, x_n$ is a basis for the dual space $V^* = \mathrm{Hom}_K(V, K)$, then

$$K[V] = K[x_1, \ldots, x_n] = K \oplus V^* \oplus S^2(V^*) \oplus S^3(V^*) \oplus \cdots$$

Here, $S^m(V^*)$ denotes the symmetric mth power of V^*, which consists of the homogeneous polynomials of degree m in $x_1, \ldots, x_n$. Thus for example $S^2(V^*)$ has a basis consisting of the monomials $x_i x_j$ for the $\binom{n+1}{2}$ choices of i and j. In general, the dimension of $S^m(V^*)$ as a vector space is $\binom{n+m-1}{m} = \binom{n+m-1}{n-1}$. We regard $K[V]$ as a graded ring by putting each x_i in degree one. (If you are a topologist, then you may wish to double all the degrees.) The group G acts on $K[V]$ via $(gf)(v) = f(g^{-1}v)$, and the basic object of study of invariant theory is the set of all fixed points of this action, the **ring of invariants** $K[V]^G$.

Warning If K is a finite field $\mathbb{F}_q$, then we should not really regard $K[V]$ as a ring of functions on V, since for example x_i and x_i^q take the same value at all points of V. Rather, we regard $K[V]$ as a ring of functions on $\bar{K} \otimes_K V$ fixed by the Galois automorphisms $\mathrm{Gal}(\bar{K}/K)$, where $\bar{K}$ is an algebraic closure of K.

If $\chi : G \to K^\times$ is a 1–dimensional representation of G, the module of χ–relative invariants, denoted by $K[V]^G_\chi$ is defined by

$$K[V]^G_\chi = \{f \in K[V] \mid g(f) = \chi(g) f \forall g \in G\}.$$

More generally, if S is a simple KG-module, we set

$$K[V]^G_S = S \otimes_{\mathrm{End}_{KG}(S)} \mathrm{Hom}_{KG}(S, K[V])$$

(note that if K is algebraically closed then $\mathrm{End}_{KG}(S) = K$). The evaluation map

$$S \otimes_{\mathrm{End}_{KG}(S)} \mathrm{Hom}_{KG}(S, K[V]) \to K[V]$$

is injective, and identifies $K[V]^G_S$ with the largest subspace of $K[V]$ consisting of a direct sum of copies of S. Multiplication in $K[V]$ gives a map

$$K[V]^G \times K[V]^G_S \to K[V]^G_S$$

making $K[V]^G_S$ into a $K[V]^G$-module.

At a coarser level, we could look at the action of G on the field of fractions $K(V)$ of $K[V]$. This may be regarded as the field of rational functions on V. The action of G is given by $g(f_1/f_2) = g(f_1)/g(f_2)$.

Recall that if we are given an extension of commutative rings $B \supseteq A$, we say that an element of B is **integral** over A if it satisfies a monic polynomial (i.e., a polynomial whose leading coefficient is one) with coefficients in A. This is the same as saying that the element lies in a subring of B which contains A and is finitely generated as an A-module. So the sum and product of integral elements is again integral. We say that B is an integral extension of A if every element of B is integral over A. If B is an integral extension of A and finitely generated over A as a ring, then it is finitely generated over A as a module. In this case, we say that B is a **finite** extension of A. Finally, we say an integral domain A is **integrally closed** if every element of the field of fractions of A which is integral over A is in A. For example, a unique factorization domain is always integrally closed.

Proposition 1.1.1 *Suppose that V is a finite dimensional faithful representation of a finite group G over a field K. Then $K(V)$ is a Galois (i.e., normal and separable) extension of $K(V)^G$ with Galois group G. The field $K(V)^G$ is the field of fractions of $K[V]^G$, and $K[V]^G$ is integrally closed in $K(V)^G$.*

Proof Since G acts as field automorphisms on $K(V)$, it is clear that $K(V)$ is a Galois extension of $K(V)^G$ with Galois group G.

Every element of $K(V)$ may be written in the form f_1/f_2 with f_2 G-invariant, just by multiplying the top and bottom by the distinct images of f_2. It follows from this that $K(V)^G$ is the field of fractions of $K[V]^G$.

Any element $f \in K(V)^G$ which is integral over $K[V]^G$ is also integral over $K[V]$. Since $K[V]$ is integrally closed in $K(V)$, this means that $f \in K[V]$. Since f is G-invariant, this means that $f \in K[V]^G$. □

The question of whether $K(V)^G$ is pure transcendental over K is a hard one in general. For a nilpotent group in coprime characteristic, the answer is affirmative, Morikawa [62]. The first examples where $K(V)^G$ is not pure transcendental with $K = \mathbb{Q}$ were produced by Swan [104], and with $K = \mathbb{C}$ by Saltman [85]. See also the survey article of Kervaire and Vust [54].

Example Let G be the symmetric group Σ_n, acting as permutations on a basis $v_1, \dots, v_n$ of V. If $x_1, \dots, x_n$ is the dual basis of V^*, then the elements of

$$K[V]^{\Sigma_n} = K[x_1, \dots, x_n]^{\Sigma_n}$$

are called **symmetric functions.** The "fundamental theorem on symmetric functions" says that they form a polynomial ring in generators $e_i(x_1, \dots, x_n)$ $(1 \le i \le n)$ called the **elementary symmetric functions** defined by

$$f(X) = \prod_{i=1}^{n}(X - x_i) = X^n + \sum_{i=1}^{n}(-1)^i e_i(x_1, \dots, x_n) X^{n-i},$$

where X is an indeterminate. To prove that $K[V]^{\Sigma_n} = K[e_1, \dots, e_n]$, we argue as follows. First of all, Σ_n acts as Galois automorphisms of $K(V)$ fixing $K(e_1, \dots, e_n)$. Now $K(V)$ is the splitting field of f over $K(e_1, \dots, e_n)$, and $df/dX \ne 0$, so the extension is Galois. Any Galois automorphism must permute the roots of $f(X)$, namely $x_1, \dots, x_n$, and is determined by this permutation, and so Σ_n is precisely the Galois group, and $K(V)^{\Sigma_n} = K(e_1, \dots, e_n)$. Now $K[V]$ is integral over $K[e_1, \dots, e_n]$, and hence so is $K[V]^{\Sigma_n} = K(V)^{\Sigma_n} \cap K[V]$. Since $K[e_1, \dots, e_n]$ is integrally closed, this completes the proof.

For a more direct proof, see Macdonald [56], §I.2.

1.2 Noetherian rings and modules

Let G be a finite group, K a field and V a finite dimensional KG-module. The fact that $K[V]^G$ is finitely generated as a K-algebra follows from the work of Hilbert in the case where $|G|$ is coprime to the characteristic of K, and was proved by Noether [78, 79] in the general case. We shall present three proofs, each with its advantages and disadvantages (Theorem 1.3.1, Corollary 1.5.3 and Theorem 1.6.3).

The first proof depends on a certain amount of commutative algebra, and we begin with some comments. Classically, one of the main motivations for the development of commutative algebra was to put algebraic geometry on a firm foundation. Two topics which provided impetus are intersection theory (various forms of Bézout's theorem and generalizations) and invariant theory, the topic of this book. In the course of this book, we shall need to make use of a considerable amount of commutative algebra, which we introduce as we need it. Good general references for commutative algebra are Matsumura [59] and Serre [90].

A module M over a commutative ring A is said to be Noetherian if every ascending chain of submodules is eventually constant. Clearly, every finite direct sum of Noetherian modules, as well as every submodule and quotient of a Noetherian module is again Noetherian. The ring A itself is said to be Noetherian if it is so as a module over itself; in other words, if every ascending chain of ideals is eventually constant.

Proposition 1.2.1 *A module M over a Noetherian ring A is Noetherian if and only if it is finitely generated.*

Proof If M is finitely generated, then it is a quotient of a finite direct sum of copies of A, and hence Noetherian. Conversely, if M is Noetherian, we choose a sequence of elements $x_1, x_2, \ldots$ in M with each x_i not in the submodule generated by $x_1, \ldots, x_{i-1}$. Since M is Noetherian, such a sequence must terminate, and so M is finitely generated. □

Corollary 1.2.2 *A submodule of a finitely generated module over a Noetherian ring is again finitely generated.* □

Lemma 1.2.3 *A commutative ring A is Noetherian if and only if every ideal is finitely generated.*

Proof If A is Noetherian then by the corollary, every ideal is finitely generated. Conversely, suppose that every ideal of A is finitely generated. Given an ascending chain of ideals of A, the union is an ideal in A, and hence finitely generated. But any finite subset of the union lies in one of the ideals, and so the chain is constant from that ideal onwards. □

Theorem 1.2.4 (Hilbert's basis theorem) *If A is a commutative Noetherian ring, then so is the polynomial ring $A[x]$.*

Proof By the lemma, it suffices to prove that if I is an ideal in $A[x]$ then I is finitely generated. Let I' be the ideal in A generated by the leading coefficients of the polynomials in I. Since A is Noetherian, I' is finitely generated, say by $a_1, \ldots, a_n$. Choose polynomials $f_1, \ldots, f_n \in I$ with these leading coefficients, let t be the maximum of the degrees of the f_i, and write I_0 for the ideal in $A[x]$ generated by $f_1, \ldots, f_n$. If f is any polynomial in I of degree at least t, then the leading coefficient of f is in I', so that we may subtract off an A-linear combination of products of f_i's with powers of x, in order to get rid of the leading coefficient and hence decrease the degree. Eventually the degree is less than t, and so we have

$$I = I_0 + (I \cap (A \oplus Ax \oplus \cdots \oplus Ax^{t-1})).$$

It follows that I is finitely generated. □

Corollary 1.2.5 *If A is a finitely generated commutative algebra over a field K, then A is Noetherian.*

Proof A is a quotient of some polynomial ring over K, which is Noetherian by the theorem. □

1.3 Finite groups in arbitrary characteristic

The first proof we present of the finite generation of the invariants is as follows.

Theorem 1.3.1 (Hilbert–Noether) *Suppose that K is a field, and G is a finite group acting as automorphisms of a finitely generated commutative K-algebra A (for example $A = K[V]$). Then A^G is also a finitely generated commutative K-algebra and A is finitely generated as a module over A^G.*

Proof If $a \in A$ then a satisfies the monic polynomial

$$\prod_{g \in G}(X - g(a)) \in A^G[X]$$

and so A is an integral extension of A^G. Let A' be the subalgebra of A^G generated by the coefficients of the monic polynomials satisfied by a finite set of K-algebra generators of A. Then A' is a finitely generated K-algebra, and hence Noetherian. A is a finitely generated A'-module, and hence so is A^G by Corollary 1.2.2. Thus A^G is a finitely generated K-algebra. □

Corollary 1.3.2 *Suppose that G is a finite group, K is a field and V is a finite dimensional KG-module. Then $K[V]^G$ is a finitely generated K-algebra. If S is a simple KG-module then $K[V]^G_S$ is a finitely generated $K[V]^G$-module.* □

1.4 Krull dimension and going up and down

Having seen that $K[V]$ is a finite extension of the finitely generated K-algebra $K[V]^G$, we can use this to describe the relationship between the prime ideals of $K[V]$ and $K[V]^G$.

We first discuss Krull dimension. The **Krull dimension** of a commutative ring A is the maximum length n of a chain of proper inclusions of prime ideals $\mathfrak{p}_0 \supset \mathfrak{p}_1 \supset \cdots \supset \mathfrak{p}_n$, or ∞ if there are such chains of unbounded length. If M is an A-module, we define the Krull dimension of M to be the Krull dimension of the ring $A/\mathrm{Ann}_A(M)$, where $\mathrm{Ann}_A(M) = \{a \in A \mid \forall m \in M,\ am = 0\}$ is the **annihilator** in A of M.

We write $\dim(A)$ and $\dim(M)$, with no subscript, to denote the Krull dimension of A and M. Whenever we wish to denote the dimension of a vector space, we keep the field as a subscript, to avoid confusion.

Proposition 1.4.1 *The Krull dimension of $K[V]$ is equal to* $\dim_K(V)$.

Proof Suppose that $K[V] = K[x_1, \ldots, x_n]$. The chain of prime ideals

$$(x_1, \ldots, x_n) \supset (x_2, \ldots, x_n) \supset \cdots \supset (x_n) \supset 0$$

shows that the Krull dimension is at least n. To prove that it is at most n, it suffices to show that if $\mathfrak{p} \subset \mathfrak{p}'$ are prime ideals then the transcendence degree over K of (the field of fractions of) $K[V]/\mathfrak{p}'$ is strictly less than that of $K[V]/\mathfrak{p}$. If this is not the case, since $K[V]/\mathfrak{p}$ surjects onto $K[V]/\mathfrak{p}'$, then the transcendence degrees are equal. Reorder the x_j so that $x_1, \ldots, x_r$ form a transcendence base for $K[V]/\mathfrak{p}'$ and hence also for $K[V]/\mathfrak{p}$. Let $K' = K(x_1, \ldots, x_r)$, so that we may extend $\mathfrak{p}$ and $\mathfrak{p}'$ to ideals $\mathfrak{P} \subset \mathfrak{P}'$ in $K'[x_{r+1}, \ldots, x_n]$. Each of $x_{r+1}, \ldots, x_n$ is algebraic in $K'[x_{r+1}, \ldots, x_n]/\mathfrak{P}$, and so this is already a field, which contradicts the proper inclusion of $\mathfrak{P}$ in $\mathfrak{P}'$. □

We relate the prime ideals in $K[V]$ and $K[V]^G$ using Theorem 1.4.4. Before discussing this, we introduce localization, which is a way of turning prime ideal into maximal ideals.

If A is an integral domain, and $\mathfrak{p}$ is a prime ideal in A, we write $A_\mathfrak{p}$ for the subring of the field of fractions of A consisting of elements xy^{-1} with $x, y \in A$ and $y \notin \mathfrak{p}$. More generally, if A is a commutative ring, we write $A_\mathfrak{p}$ for the ring whose elements are expressions of the form x/y with $x, y \in A$ and $y \notin \mathfrak{p}$. We regard x/y and x'/y' as equal if for some $z \in A$, $z \notin \mathfrak{p}$ we have $z(xy' - x'y) = 0$.

If M is an A-module then $M_\mathfrak{p} = A_\mathfrak{p} \otimes_A M$ is an $A_\mathfrak{p}$-module. The kernel of the map $M \to M_\mathfrak{p}$ consists of the elements which are annihilated by some element of A not in $\mathfrak{p}$. A short exact sequence $0 \to M_1 \to M_2 \to M_3 \to 0$ of A-modules gives rise to a short exact sequence

$$0 \to (M_1)_\mathfrak{p} \to (M_2)_\mathfrak{p} \to (M_3)_\mathfrak{p} \to 0$$

of $A_\mathfrak{p}$-modules (i.e., $A_\mathfrak{p}$ is **flat** over A). To see this, it suffices to see that the first map is injective. This is clear, because whether an element of M_1 is annihilated by an element of A not in $\mathfrak{p}$ does not depend on whether we view it as an element of M_1 or M_2.

If I is an ideal in $A_\mathfrak{p}$ then $I \cap A$ is an ideal in A contained in $\mathfrak{p}$ (because if it contains an element outside $\mathfrak{p}$ then it contains the identity element). Conversely, if I' is an ideal in A contained in $\mathfrak{p}$ then $I'_\mathfrak{p}$ is an ideal in $A_\mathfrak{p}$. These processes set up a one–one correspondence between ideals in $A_\mathfrak{p}$ and ideals in A contained in $\mathfrak{p}$. In particular, $\mathfrak{p}_\mathfrak{p}$ is the unique maximal ideal of $A_\mathfrak{p}$. A commutative ring with a unique maximal ideal is called a **local ring**.

Lemma 1.4.2 *Suppose that $B \supseteq A$ is a finite extension of commutative rings. If $\mathfrak{P}$ is a maximal ideal of B then $\mathfrak{P} \cap A$ is a maximal ideal of A. Conversely if $\mathfrak{p}$ is a maximal ideal of A then there is a prime ideal $\mathfrak{P}$ of B with $\mathfrak{P} \cap A = \mathfrak{p}$, and any such prime ideal $\mathfrak{P}$ is maximal.*

Proof If $\mathfrak{P}$ is maximal in B then $B/\mathfrak{P}$ is a field. It is integral over $A/(\mathfrak{P}\cap A)$, so if $x \in A/(\mathfrak{P}\cap A)$ then $x^{-1} \in B/\mathfrak{P}$ satisfies some monic equation over $A/(\mathfrak{P}\cap A)$, say $x^{-n}+a_{n-1}x^{-n+1}+\cdots+a_0 = 0$. But then $x^{-1} = -a_{n-1}-\cdots-a_0x^{n-1} \in A/(\mathfrak{P}\cap A)$, so $A/(\mathfrak{P}\cap A)$ is a field and $\mathfrak{P}\cap A$ is maximal. Conversely, if $\mathfrak{P}\cap A$ is maximal then $B/\mathfrak{P}$ is an integral extension of the field $A/(\mathfrak{P}\cap A)$, and is hence a field, so that $\mathfrak{P}$ is maximal.

If $\mathfrak{p}$ is a maximal ideal in A, suppose that $\mathfrak{p}B = B$. If B is generated over A by $b_1,\dots,b_n$ then we have $b_i = \sum_j x_{ij}b_j$ with $x_{ij} \in \mathfrak{p}$. But then $\det(I-(x_{ij})) \in A$ annihilates each b_i and hence annihilates $1 \in B$, so this determinant is zero. But it is congruent to 1 modulo $\mathfrak{p}$. This contradiction shows that $\mathfrak{p}B \neq B$, so $\mathfrak{p}B$ is contained in some maximal ideal of B, which therefore intersects A in $\mathfrak{p}$. □

Lemma 1.4.3 *If an ideal I is contained in a finite union of primes $I \subseteq \mathfrak{p}_1\cup\cdots\cup\mathfrak{p}_n$ then I is contained in some $\mathfrak{p}_i$.*

Proof Suppose n is minimal, so that I is not contained in the union of any proper subset of $\{\mathfrak{p}_1,\dots,\mathfrak{p}_n\}$, and suppose that $n > 1$. Choose $y \in I$ with $y \notin \mathfrak{p}_1\cup\cdots\cup\mathfrak{p}_{n-1}$. Then $y \in \mathfrak{p}_n$. Choose $z \in I$ with $z \notin \mathfrak{p}_n$, and $t_i \in \mathfrak{p}_i$ with $t_i \notin \mathfrak{p}_n$ for each i between 1 and $n-1$. Then $y+zt_1\dots t_{n-1}$ is in I but is not in $\mathfrak{p}_1\cup\cdots\cup\mathfrak{p}_n$. This contradiction proves the lemma. □

Remark In the above proof, we only used the fact that $\mathfrak{p}_n$ was prime, and even this is only necessary if $n > 2$. So in fact it suffices to assume that all except possibly two of the $\mathfrak{p}_i$ are prime.

Theorem 1.4.4 *Suppose that $B \supseteq A$ is a finite extension of commutative rings.*

(i) **(Lying over)** *If $\mathfrak{p}$ is a prime ideal of A then there is a prime ideal $\mathfrak{P}$ of B with $\mathfrak{P}\cap A = \mathfrak{p}$. There are no strict inclusions between such prime ideals $\mathfrak{P}$. In this situation we say that $\mathfrak{P}$* **lies over** *$\mathfrak{p}$.*

(ii) **(Going up)** *If $\mathfrak{p}' \supset \mathfrak{p}$ are prime ideals in A and $\mathfrak{P}$ is a prime ideal in B lying over $\mathfrak{p}$ then there is a prime ideal $\mathfrak{P}'$ in B lying over $\mathfrak{p}'$ with $\mathfrak{P}' \supset \mathfrak{P}$.*

(iii) **(Transitivity)** *Suppose that A and B are integrally closed domains, and that the corresponding extension $L' \supseteq L$ of fields of fractions is* **normal** *(i.e., an irreducible polynomial over L with a root in L' splits completely in L', but is not necessarily separable). The Galois group $G = \mathrm{Gal}(L'/L)$ acts transitively on the prime ideals $\mathfrak{P}$ of B lying over the given prime ideal $\mathfrak{p}$ of A.*

(iv) **(Going down)** *Suppose that A and B are integral domains, and A is integrally closed. If $\mathfrak{p} \subset \mathfrak{p}'$ are prime ideals in A and $\mathfrak{P}'$ is a prime ideal of B lying over $\mathfrak{p}'$ then there is a prime ideal $\mathfrak{P}$ in B lying over $\mathfrak{p}$, with $\mathfrak{P} \subset \mathfrak{P}'$.*

Proof (i) We form the localization $A_\mathfrak{p}$ of A at $\mathfrak{p}$ and set $B_\mathfrak{p} = A_\mathfrak{p} \otimes_A B$. Then $B_\mathfrak{p}$ is finite over $A_\mathfrak{p}$. The prime ideals $\mathfrak{P}$ in B with $\mathfrak{P}\cap A = \mathfrak{p}$ correspond to the prime ideals $\mathfrak{P}_\mathfrak{p}$ in $B_\mathfrak{p}$ with $\mathfrak{P}_\mathfrak{p} \cap A_\mathfrak{p} = \mathfrak{p}_\mathfrak{p}$. So it is enough to consider the case where $\mathfrak{p}$ is maximal, and this was dealt with in Lemma 1.4.2.

(ii) To prove this, pass down to the quotient $B/\mathfrak{P} \supseteq A/\mathfrak{p}$ and use (i).

(iii) First we treat the case where $L' \supseteq L$ is separable, so that $L = (L')^G$ and $A = B^G$ (by integral closure). Suppose that $\mathfrak{P}$ and $\mathfrak{P}'$ are prime ideals in B lying over $\mathfrak{p}$, and suppose that $\mathfrak{P}$ and $\mathfrak{P}'$ are not G-conjugate. By part (i), no conjugate of $\mathfrak{P}$ contains $\mathfrak{P}'$. So by Lemma 1.4.3, we may choose an element $x \in \mathfrak{P}'$ such that for all $g \in G$, $x \notin g(\mathfrak{P})$. Then $\prod_{g \in G} g(x)$ is an element of $B^G = A$ lying in $\mathfrak{P}'$ but not in $\mathfrak{P}$, which contradicts the hypothesis that $\mathfrak{P}' \cap A = \mathfrak{P} \cap A = \mathfrak{p}$.

Next, we treat the case where $L' \supseteq L$ is purely inseparable. In this case, the only prime ideal of B lying over $\mathfrak{p}$ is $\{x \in B \mid x^{p^n} \in \mathfrak{p} \text{ for some } n \geq 0\}$, where p is the characteristic of L. Finally, every extension is a composition of a purely inseparable and a separable extension, so the statement is proved.

(iv) Let $L' \supseteq L$ be the fields of fractions of $B \supseteq A$, and let L'' be a finite normal extension of L containing L'. Let C be the integral closure of A in L'', so that C is a finite extension of B, and choose primes $\mathfrak{Q}$ and $\mathfrak{Q}'$ in C lying over $\mathfrak{p}$ in A and $\mathfrak{P}'$ in B respectively. By (ii), we can find $\mathfrak{Q}'' \supset \mathfrak{Q}$ with $\mathfrak{Q}''$ lying over $\mathfrak{p}'$. By (iii), for some $g \in \mathrm{Gal}(L''/L)$ we have $g(\mathfrak{Q}'') = \mathfrak{Q}'$. Set $\mathfrak{P} = g(\mathfrak{Q}) \cap A$. Then $\mathfrak{P} \subset g(\mathfrak{Q}'') \cap B = \mathfrak{Q}' \cap B = \mathfrak{P}'$ and $\mathfrak{P}$ lies over $\mathfrak{p}$. □

Corollary 1.4.5 *If $B \supseteq A$ is a finite extension of commutative rings, then the Krull dimensions of A and B are equal.* □

Corollary 1.4.6 *Suppose that G is a finite group, K is a field and V is a finite dimensional KG-module. Then the Krull dimension of $K[V]^G$ is equal to* $\dim_K(V)$.

Proof Apply the previous corollary to the extension $K[V] \supseteq K[V]^G$, and use Proposition 1.4.1. □

Remark According to Hilbert's weak Nullstellensatz (see Matsumura [59], Theorem 5.3) if K is algebraically closed then the maximal ideals in $K[V]$ are in natural one–one correspondence with the points of V, $V \cong \max K[V]$. By part (i) of the theorem, the inclusion $i : K[V]^G \to K[V]$ gives rise to a *surjective* map

$$i^* : \max K[V] \to \max K[V]^G,$$

and by part (iii) of the theorem we have

$$\max (K[V]^G) \cong V/G.$$

1.5 Noether's bound in characteristic zero

Noether [78] proved that if K is a field of characteristic zero, then $K[V]^G$ is generated by elements of degree at most $|G|$. In particular, the number of generators needed is at most

$$\binom{\dim_K(V)+|G|}{|G|}.$$

In fact, this is a special case of a relative version, which says that if H is a subgroup of G and $K[V]^H$ is generated by elements of degree at most m, then $K[V]^G$ is generated by elements of degree at most $m.|G:H|$. The proof we give is a relativization, due to B. Schmid [88], of an argument of G. W. Schwarz and R. P. Stanley.

Before we begin, we introduce the transfer. If H is a subgroup of G and V is a finite dimensional KG-module, we define

$$\mathrm{Tr}_{H,G} : K[V]^H \to K[V]^G$$

by

$$\mathrm{Tr}_{H,G}(f)(x) = \sum_{g\in G/H} g(f)(x) = \sum_{g\in G/H} f(g^{-1}(x)).$$

The notation means that the sum runs over a set of left coset representatives of H in G. Note that the composite

$$K[V]^G \hookrightarrow K[V]^H \xrightarrow{\mathrm{Tr}_{H,G}} K[V]^G$$

is equal to multiplication by $|G:H|$. In particular, if $|G:H|$ is invertible in K then the transfer is surjective, and the map

$$\pi_{G,H} = \frac{1}{|G:H|}\mathrm{Tr}_{H,G} : K[V]^H \to K[V]^G \hookrightarrow K[V]^H$$

is an idempotent projection whose image is equal to $K[V]^G$. In particular, $K[V]^H$ is a direct sum of $K[V]^G$ and the kernel of $\pi_{G,H}$. If $H=1$, we just write π_G for $\pi_{G,1} = \frac{1}{|G|}\sum_{g\in G} g$, so that $K[V] = K[V]^G \oplus \ker(\pi_G)$.

We shall also need the following combinatorial lemma.

Lemma 1.5.1 *In characteristic zero, every monomial $u_1 \dots u_j$ is a linear combination of jth powers of sums of subsets of $u_1, \dots, u_j$.*

Proof This follows from the formula

$$j!\,u_1 \dots u_j = \sum_{I\subseteq\{1,\dots,j\}} (-1)^I \left(\sum_{i\in I} u_i\right)^j.$$

In this formula, I runs over all subsets of $\{1,\dots,j\}$. □

Theorem 1.5.2 *Suppose that V is a finite dimensional representation of a finite group G over a field K of characteristic zero. If H is a subgroup of G and $K[V]^H$ is generated by elements of degree at most m, then $K[V]^G$ is generated by elements of degree at most $m.|G : H|$.*

Proof Let A be the subalgebra of $K[V]^G$ generated by elements of degree at most md, where $d = |G : H|$, and let M be the linear subspace of $K[V]^H$ spanned by elements of degree at most $md - 1$. First, we claim that $K[V]^H = AM$. Elements of degree at most $md - 1$ are clearly in AM. Every element of $K[V]^H$ is a linear combination of products of elements of degree at most m, so it suffices, by the lemma, to show that for u a linear combination of elements of $K[V]^H$ of degree at most m, all the powers u^j are in AM. For $j < d$, this is clear. Now u is a root of the equation

$$\prod_{g \in G/H} (X - g(u)) = X^d + a_{d-1}X^{d-1} + \cdots + a_0$$

(the product is over a set of coset representatives of H in G) with coefficients $a_i \in A$, and so $u^d = -a_{d-1}u^{d-1} - \cdots - a_0 \in AM$. Then we have $u^{d+1} = -a_{d-1}u^d - \cdots - a_0 u \in AM$, and continuing by induction, all u^j are in AM, which completes the proof that $K[V]^H = AM$.

Finally, we apply the projection $\pi_{G,H}$ described above, to obtain

$$K[V]^G = \pi_{G,H}(K[V]^H) = \pi_{G,H}(AM) = A\pi_{G,H}(M) = A.$$

□

Corollary 1.5.3 (Noether) *Suppose that V is a finite dimensional representation of a finite group G over a field K of characteristic zero. Then $K[V]^G$ is generated by elements of degree at most $|G|$.* □

1.6 Linearly reductive algebraic groups

The third proof of finite generation that we offer also has the disadvantage of not working in the modular case, but has the advantage that it generalizes to a suitable class of linear algebraic groups. We say that a subgroup of $GL_n(K)$ is a **linear algebraic group** if it is closed in the Zariski topology, i.e., it can be written as the set of zeros of a collection of polynomials in the n^2 variables. Note that $GL_n(K)$ is the Zariski open subset of $\mathrm{Mat}_n(K) \cong K^{n^2}$ given by the non-vanishing of the determinant. By adding an extra coordinate corresponding to the inverse of the determinant, one can consider $GL_n(K)$ as a closed subgroup of K^{n^2+1}. Examples of linear algebraic groups are given by $GL_n(K)$, $SL_n(K)$, $Sp_{2n}(K)$, the unitary group $U(n)$, the group of diagonal matrices T^n, the group of upper triangular matrices with ones on the diagonal $\mathrm{Uni}(n)$, any finite group, and so on.

A **rational representation** of a linear algebraic group is a homomorphism $G \to GL_m(K)$ with the property that the m^2 entries in the matrices are polynomials in the n^2+1 variables given by the matrix entries of G and the inverse of the determinant. The inclusion of this last variable enables us to regard the dual of a rational representation as a rational representation.

In this generality, finite generation no longer holds for $K[V]^G$, even in case $K = \mathbb{C}$. Nagata [65] (1959) showed that if G is a subgroup of $\mathrm{Uni}(2) \times \cdots \times \mathrm{Uni}(2)$ (n^2 copies, $n \geq 4$, regarded as a subgroup of $GL_{2n^2}(\mathbb{C})$) described by 3 linear relations "in general position" then $\mathbb{C}[V]^G$ is not finitely generated. See also Dieudonné and Carrell [32], and Pommerening [82].

A linear algebraic group is said to be **linearly reductive** if every finite dimensional representation is completely reducible (i.e., a direct sum of irreducible representations).

Proposition 1.6.1 *A linear algebraic group G is linearly reductive if and only if for every rational representation V of G there is a G-invariant K-linear operator $\pi : V \to V$ which projects V onto the G-invariants. Such an operator π is necessarily unique.*

Proof We first show that if π exists, then it is unique. If π_1 and π_2 are two such maps then $\pi_1(1-\pi_1)$ is zero, so that the image of $1-\pi_1$ contains no non-zero G-invariants. Thus $\pi_2(1-\pi_1)$ is also zero. Furthermore, π_2 acts as the identity on the image of π_1, so we have $\pi_1 = \pi_2\pi_1 = \pi_2$.

We now turn to the equivalence of the two statements. One direction is clear. Namely, for a linearly reductive group there is such a map π. For the other direction, first note that because of the uniqueness of π, if $\alpha : V \to W$ is a homomorphism of representations then $\pi_W \circ \alpha = \alpha \circ \pi_V$, because both are the projection of $\mathrm{Im}(\alpha)$ onto its G-invariants. In particular, if $\alpha : V \to W$ is surjective, then so is the map $\alpha|_{V^G} : V^G \to W^G$.

Now suppose that $V' \leq V$ is G-invariant. We make $\mathrm{Hom}_K(V', V)$ into a rational representation in the normal way: $g(\phi)(v') = g(\phi(g^{-1}v'))$. The injection $i : V' \hookrightarrow V$ is a G-invariant element of $\mathrm{Hom}_K(V', V)$. Consider composition of maps

$$i^* : \mathrm{Hom}_K(V, V') \to \mathrm{Hom}_K(V', V').$$

This is surjective, and so the identity element is the image of an invariant element $\rho \in \mathrm{Hom}_K(V, V')$. The kernel of ρ is then an invariant complement to V' in V. □

As an example, if G is the complexification of a compact real Lie group G_0 (for example, $GL_n(\mathbb{C})$ is the complexification of $U(n)$) then we can take

$$\pi(x) = \int_{G_0} g(x)dg$$

where the integral is with respect to normalized Haar measure on G_0, and so G is linearly reductive. This is called Weyl's **unitarian trick.** It turns out that every linearly reductive group over $\mathbf{C}$ has this form.

As a particular case of this, if G is finite then the above map is $\pi = \pi_G$ (Section 1.5), and this shows that G is linearly reductive. The same argument works over any field of characteristic coprime to $|G|$.

Lemma 1.6.2 *If G is linearly reductive then the map $\pi : K[V] \to K[V]$ is a $K[V]^G$-module homomorphism.*

Proof If $x \in K[V]^G$ then the maps $y \mapsto x\pi(y)$ and $y \mapsto \pi(xy)$ are both projections of $x.K[V]$ onto the invariants $x.K[V]^G = (x.K[V])^G$, so by uniqueness we have $x\pi(y) = \pi(xy)$. □

Theorem 1.6.3 *If G is a linearly reductive algebraic group then $K[V]^G$ is a finitely generated K-algebra.*

Proof Let I be the ideal in $K[G]$ generated by the homogeneous invariants of positive degree. By Hilbert's basis theorem, $K[G]$ is Noetherian, and so I may be generated as an ideal by a finite number of invariant elements $f_1, \ldots, f_s$. We claim that the invariants are generated as a K-algebra by $f_1, \ldots, f_s$. If f is any homogeneous invariant of positive degree, then $f \in I$, so we can write $f = \sum \alpha_i f_i$ with $\alpha_i \in K[V]$. By the lemma, when we apply π we get

$$f = \sum \pi(\alpha_i) f_i.$$

The $\pi(\alpha_i)$ are of lower degree than f, so by induction we may assume that they are in the subalgebra generated by $f_1, \ldots, f_s$, and hence so is f. □

Chapter 2

Poincaré series

2.1 The Hilbert–Serre theorem

Let $G \to GL(V)$ be a linear representation of the finite group G. We would like to know the dimension of the homogeneous component $K[V]^G_j$ consisting of invariant polynomials of degree j. We encode this information in terms of the Poincaré series

$$p(K[V]^G, t) = \sum_{j=0}^{\infty} t^j \dim_K K[V]^G_j.$$

Taking $A = M = K[V]^G$ in the following theorem, we see that one of the consequences of finite generation is that this power series is a rational function of t whose poles are at roots of unity.

Theorem 2.1.1 (Hilbert, Serre) *Suppose that $A = \bigoplus_{j=0}^{\infty} A_j$ is a commutative graded ring with $A_0 = K$, and finitely generated over K by homogeneous elements $x_1, \ldots, x_s$ in positive degrees $k_1, \ldots, k_s$. Suppose that $M = \bigoplus_{j=-\infty}^{\infty} M_j$ is a finitely generated graded A-module (i.e., we have $A_i.A_j \subseteq A_{i+j}$ and $A_i.M_j \subseteq M_{i+j}$, and $M_i = 0$ for all but a finite number of negative values of i). Then the Poincaré series $p(M,t)$ is of the form*

$$\frac{f(t)}{\prod_{j=1}^{s}(1 - t^{k_j})}$$

where $f(t)$ is a Laurent polynomial in t with integer coefficients.

Proof We work by induction on s. If $s = 0$ then $p(M,t)$ is a Laurent polynomial, so suppose $s > 0$. Denoting by M' and M'' the kernel and cokernel of multiplication by x_s, we have an exact sequence

$$0 \to M'_r \to M_r \xrightarrow{x_s} M_{r+k_s} \to M''_{r+k_s} \to 0.$$

Now M' and M'' are finitely generated graded modules for $K[x_1, \ldots, x_{s-1}]$, and so by induction their Poincaré series have the given form. From the above exact sequence we have

$$t^{k_s} p(M', t) - t^{k_s} p(M, t) + p(M, t) - p(M'', t) = 0.$$

Thus

$$p(M, t) = \frac{p(M'', t) - t^{k_s} p(M', t)}{1 - t^{k_s}}$$

has the given form. □

In this situation, the following proposition shows that we may interpret the order of the pole at $t = 1$ as telling us about the polynomial rate of growth of the graded pieces.

Proposition 2.1.2 *Suppose that*

$$p(t) = \frac{f(t)}{\prod_{j=1}^{s}(1 - t^{k_j})} = \sum_{r \geq 0} a_r t^r$$

where $f(t)$ is a Laurent polynomial with integer coefficients and the a_r are non-negative integers. Let γ be the order of the pole of $p(t)$ at $t = 1$. Then

(i) *there exists a constant $\kappa > 0$ such that $a_n \leq \kappa . n^{\gamma - 1}$ for $n > 0$, but*

(ii) *if $\gamma \geq 1$, there does not exist a constant $\kappa > 0$ such that $a_n \leq \kappa . n^{\gamma - 2}$ for $n > 0$.*

Proof The hypothesis and conclusion remain unaltered if we replace $p(t)$ by $p(t).(1 + t + \cdots + t^{k_j - 1})$, and so without loss of generality each $k_j = 1$. So we may suppose $p(t) = f(t)/(1-t)^\gamma$ with $f(1) \neq 0$. We may also multiply $p(t)$ by some positive power of t if necessary in order to make $f(t)$ a polynomial, without interfering with the conclusion. So suppose that $f(t) = \alpha_m t^m + \cdots + \alpha_0$. We have

$$a_n = \alpha_0 \binom{n + \gamma - 1}{\gamma - 1} + \alpha_1 \binom{n + \gamma - 2}{\gamma - 1} + \cdots + \alpha_m \binom{n + \gamma - m - 1}{\gamma - 1}.$$

The condition $f(1) \neq 0$ implies that $\alpha_0 + \cdots + \alpha_m \neq 0$, so this expression is a polynomial of degree exactly $\gamma - 1$ in n. □

Notation We write $\gamma(M)$ for the order of the pole at $t = 1$ of $p(M, t)$.

Lemma 2.1.3 *Suppose that $B \supseteq A$ is a finite extension of graded rings satisfying the conditions of Theorem 2.1.1. Then $\gamma(B) = \gamma(A)$.*

Proof As an A-module, B is finitely generated, so it is a quotient of a finitely generated free graded A-module M (i.e., a finite direct sum of copies of A, possibly shifted in degree). Then $p(M,t) = f(t)p(A,t)$ for some polynomial $f(t)$ with $f(1) > 0$. Thus as power series we have

$$p(A,t) \leq p(B,t) \leq f(t)p(A,t),$$

(i.e., these inequalities hold for each coefficient of these power series). The orders of the poles at $t = 1$ of $p(A,t)$ and $f(t)p(A,t)$ are equal, and so it follows from Proposition 2.1.2 that $\gamma(A) = \gamma(B)$. □

Proposition 2.1.4 $\gamma(K[V]^G) = \dim_K(V)$.

Proof By Theorem 1.3.1, $K[V]$ is a finite extension of $K[V]^G$, so by the lemma, it suffices to look at the pole at $t = 1$ of $p(K[V],t) = 1/(1-t)^n$, where $n = \dim_K(V)$. □

In Section 1.4, we saw that the Krull dimension of $K[V]^G$ is equal to $\dim_K(V)$. In fact, in the more general situation of Theorem 2.1.1, $\gamma(M)$ is equal to the Krull dimension of M. To see this, we need a graded version of Noether's normalization lemma, which is the topic of the next section. This will be a valuable tool to us later, in Chapter 4, when we discuss homological properties of $K[V]^G$.

2.2 Noether normalization

We begin with a discussion of associated primes. Suppose that A is a Noetherian commutative ring, and that M is a finitely generated A-module. We say that a prime ideal $\mathfrak{p}$ of A is an **associated prime** of M if there is a non-zero element $x \in M$ whose annihilator is exactly $\mathfrak{p}$; in other words if the map $A \to M$ sending a to ax has image isomorphic to $A/\mathfrak{p}$. We write $\mathrm{Ass}_A(M)$ for the set of associated primes of M, so that $\mathfrak{p} \in \mathrm{Ass}_A(M)$ if and only if $A/\mathfrak{p}$ is isomorphic to a submodule of M. The following lemma shows that there are "enough" associated primes.

Lemma 2.2.1 *Every maximal element (under inclusion) of the set of annihilators in A of non-zero elements of M is an associated prime. In particular, every non-zero module has at least one associated prime.*

Proof If $0 \neq x \in M$ and $\mathrm{Ann}_A(x)$ is maximal, suppose $ab \in \mathrm{Ann}_A(M)$, but $b \notin \mathrm{Ann}_A(x)$. Then $bx \neq 0$, so $\mathrm{Ann}_A(bx) \supseteq \mathrm{Ann}_A(x)$. By maximality, we have equality and so $a \in \mathrm{Ann}_A(bx) = \mathrm{Ann}_A(x)$. So $\mathrm{Ann}_A(x)$ is prime. □

Example $\mathrm{Ass}_A(A/\mathfrak{p}) = \{\mathfrak{p}\}$, since $\mathfrak{p}$ is the annihilator of any non-zero element of $A/\mathfrak{p}$.

Lemma 2.2.2 *Any finitely generated A-module M has a filtration*

$$0 = M_0 \subseteq M_1 \subseteq \cdots \subseteq M_n = M$$

with $M_i/M_{i-1} \cong A/\mathfrak{p}_i$ *for some prime* $\mathfrak{p}_i$.

Proof If $M = 0$, there is nothing to do. Otherwise, choose $\mathfrak{p}_1 \in \mathrm{Ass}_A(M)$ and let M_1 be a submodule of M isomorphic to $A/\mathfrak{p}_1$. If $M_1 \neq M$, we pass down to M/M_1 and carry on going. Since M is Noetherian, the process eventually terminates. □

Proposition 2.2.3 *A finitely generated A-module M has Krull dimension zero if and only if it has finite (composition) length.*

Proof If M has Krull dimension zero, then so does each M_i/M_{i-1} in the statement of Lemma 2.2.2, and so each $\mathfrak{p}_i$ is maximal and M has finite length.

Conversely, if M has finite length, then so does $A/\mathrm{Ann}_A(M)$. So $A/\mathrm{Ann}_A(M)$ has a finite number of maximal ideals $\mathfrak{p}_1, \ldots, \mathfrak{p}_n$, since if $\mathfrak{p}_1, \mathfrak{p}_2, \ldots$ are maximal ideals then $\mathfrak{p}_1 \supset \mathfrak{p}_1\mathfrak{p}_2 \supset \cdots$ is a descending chain of ideals. Furthermore, $\mathfrak{p}_1 \ldots \mathfrak{p}_n$ is nilpotent, so if $\mathfrak{p}$ is any prime ideal then $\mathfrak{p} \supseteq 0 = (\mathfrak{p}_1 \ldots \mathfrak{p}_n)^r$ and so $\mathfrak{p}$ contains some $\mathfrak{p}_i$, and is hence equal to it. □

Lemma 2.2.4 *If* $0 \to M_1 \to M_2 \to M_3 \to 0$ *is a short exact sequence of A-modules then*

(i) $\mathrm{Ann}_A(M_1).\mathrm{Ann}_A(M_3) \subseteq \mathrm{Ann}_A(M_2) \subseteq \mathrm{Ann}_A(M_1) \cap \mathrm{Ann}_A(M_3)$, *and*

(ii) $\mathrm{Ass}_A(M_1) \subseteq \mathrm{Ass}_A(M_2) \subseteq \mathrm{Ass}_A(M_1) \cup \mathrm{Ass}_A(M_3)$.

Proof (i) This is clear.

(ii) Since $\mathfrak{p} \in \mathrm{Ass}_A(M)$ if and only if $A/\mathfrak{p}$ is isomorphic to a submodule of M, we have $\mathrm{Ass}_A(M_1) \subseteq \mathrm{Ass}_A(M_2)$. If $\mathfrak{p} \in \mathrm{Ass}_A(M_2)$, let M' be a submodule of M_2 isomorphic to $A/\mathfrak{p}$. If $M' \cap M_1 \neq 0$ then $\mathfrak{p}$ is the annihilator of any non-zero element of $M' \cap M_1$ and so $\mathfrak{p} \in \mathrm{Ass}_A(M_1)$. If $M' \cap M_1 = 0$ then the image of M' in M_3 is isomorphic to $A/\mathfrak{p}$ and so $\mathfrak{p} \in \mathrm{Ass}_A(M_3)$. □

Proposition 2.2.5 (i) $\mathrm{Ass}_A(M)$ *is a finite set.*

(ii) *The minimal elements of* $\mathrm{Ass}_A(M)$ *are precisely the minimal primes containing* $\mathrm{Ann}_A(M)$.

Proof (i) This follows immediately from the example and Lemmas 2.2.2 and 2.2.4.

(ii) Clearly every element of $\mathrm{Ass}_A(M)$ contains $\mathrm{Ann}_A(M)$. Conversely, if $\mathfrak{p}$ is a minimal prime containing $\mathrm{Ann}_A(M)$, choose i minimal such that $\mathfrak{p}$ contains $\mathrm{Ann}_A(M_i)$ (M_i as in Lemma 2.2.2). Then by part (i) of Lemma 2.2.4, $\mathrm{Ann}_A(M_{i-1}) \subseteq$

$\mathfrak{p}_1 \cap \cdots \cap \mathfrak{p}_{i-1}$, so that $\mathfrak{p}_1 \cap \cdots \cap \mathfrak{p}_{i-1} \not\subseteq \mathfrak{p}$, while $\mathfrak{p}_1 \ldots \mathfrak{p}_i \subseteq \mathrm{Ann}_A(M_i) \subseteq \mathfrak{p}$, so that $\mathfrak{p}_i \subseteq \mathfrak{p}$. Therefore we can choose $x \in \mathfrak{p}_1 \ldots \mathfrak{p}_{i-1}$ with $x \not\in \mathfrak{p}_i$. If m is any element of M_i not in M_{i-1} then $xm \not\in M_{i-1}$, and the annihilator of xm is exactly $\mathfrak{p}_i$, so that $\mathfrak{p}_i \in \mathrm{Ass}_A(M_i) \subseteq \mathrm{Ass}_A(M)$. By minimality of $\mathfrak{p}$ we have $\mathfrak{p}_i = \mathfrak{p}$. □

Corollary 2.2.6 *The Krull dimension of M is equal to the maximum over $\mathfrak{p} \in \mathrm{Ass}_A(M)$ of $\dim(A/\mathfrak{p})$.* □

Theorem 2.2.7 (Noether normalization, graded case)
Suppose that $A = \bigoplus_{j=0}^{\infty} A_j$ is a commutative graded ring with $A_0 = K$, and finitely generated over K by homogeneous elements $x_1, \ldots, x_s$ of positive degree. Suppose that $M = \bigoplus_{j=-\infty}^{\infty} M_j$ is a finitely generated graded A-module. Then there exist homogeneous elements $f_1, \ldots, f_n$ of positive degree in A, which generate a polynomial subring $K[f_1, \ldots, f_n]$ in $A/\mathrm{Ann}_A(M)$, over which M is finitely generated as a module. The number n is equal to $\gamma(M)$, and it is also equal to the Krull dimension of M.

If $A = M$ is an integral domain, then the number n is also equal to the transcendence degree over K of (the field of fractions of) A.

Proof We note that the statement that M is finitely generated over the subring generated by $f_1, \ldots, f_n$ is equivalent to the statement that $M/(f_1, \ldots, f_n)M$ has finite dimension over K. If M has finite dimension over K, we can take $n = 0$. So we assume that this is not the case, so that $\gamma(M) \geq 1$.

Since the annihilator of an element of M annihilates each graded part, the associated primes of M are homogeneous prime ideals in A. By Proposition 2.2.5, we may choose a homogeneous element f_1 in A which is not in any associated prime of M except possibly the ideal A^+ consisting of all elements of positive degree. So the kernel of f_1 on M is of finite dimension over K, and the sequence

$$0 \to \ker(f_1 \text{ on } M) \to M \xrightarrow{f_1} M \to M/f_1M \to 0$$

shows that $\gamma(M/f_1M) = \gamma(M) - 1$. If we suppose by induction on $n = \gamma(M)$ that $f_2, \ldots, f_n$ have been chosen for M/f_1M with the desired property, then $f_1, \ldots, f_n$ have the desired property for M. They are algebraically independent, since otherwise

$$n = \gamma(M) \leq \gamma(K[f_1, \ldots, f_n]) < n.$$

By Theorem 1.4.4, the Krull dimension of M is equal to the Krull dimension of $K[f_1, \ldots, f_n]$, which by Proposition 1.4.1 is equal to n. Finally, the statement about transcendence degree is clear. □

2.3 Systems of parameters

In this section, we discuss the analogue for local rings of Noether normalization. At this stage, the reader intent on learning invariant theory may feel that this discussion is of academic interest, and safely pass on to the next section. But we should point out that this discussion becomes indispensable later on, when we prove theorems about graded rings by passing to the associated local rings. We begin this section with the opposite process.

Suppose that A is a commutative Noetherian local ring with maximal ideal $\mathfrak{M}$, so that $K = A/\mathfrak{M}$ is a field, and suppose that M is a finitely generated A-module. We set

$$\mathrm{Gr}(A) = \bigoplus_{j=0}^{\infty} \mathfrak{M}^j/\mathfrak{M}^{j+1} \qquad \mathrm{Gr}(M) = \bigoplus_{j=0}^{\infty} \mathfrak{M}^j M/\mathfrak{M}^{j+1}M.$$

Since $\mathfrak{M}^j.\mathfrak{M}^k \subseteq \mathfrak{M}^{j+k}$ and $\mathfrak{M}^j.\mathfrak{M}^k M \subseteq \mathfrak{M}^{j+k}M$, we may regard $\mathrm{Gr}(A)$ and $\mathrm{Gr}(M)$ as a graded ring and a graded module, called the **associated graded** of A and M. Since $\mathfrak{M}$ is finitely generated as an ideal, $\mathrm{Gr}(A)$ is finitely generated as a K-algebra by degree one elements, and $\mathrm{Gr}(M)$ is finitely generated over $\mathrm{Gr}(A)$ by degree zero elements.

By Proposition 2.2.3, the Poincaré series

$$p(\mathrm{Gr}(M),t) = \sum_{j=0}^{\infty} t^j \dim_K \mathfrak{M}^j M/\mathfrak{M}^{j+1}M$$

is a polynomial if and only if the Krull dimension of M is zero. If M has Krull dimension greater than zero then it has at least one non-maximal associated prime. We choose an element $x \in \mathfrak{M}$, not in $\mathfrak{M}^2$ or any non-maximal associated prime (cf. Lemma 1.4.3 and the remark following it). The Krull dimension of M/xM is then strictly smaller than that of M by Proposition 2.2.5 (ii). So we may continue this way to obtain elements $x_1, \dots, x_n$ with $n \le \dim(M)$, such that $M/(x_1, \dots, x_n)M$ has finite length. Such a sequence $x_1, \dots, x_n$ is called a **system of parameters** for M. Applying (the proof of) Theorem 2.1.1, we see that

$$p(\mathrm{Gr}(M),t) = f(t)/(1-t)^n$$

where $f(t)$ is a polynomial in t with integer coefficients, so that $p(\mathrm{Gr}(M),t)$ is a rational function of t with a pole of order $\gamma(\mathrm{Gr}(M)) \le n$ at $t = 1$. By Proposition 2.1.2, this tells us about the polynomial rate of growth of $\mathfrak{M}^j M/\mathfrak{M}^{j+1}M$.

Lemma 2.3.1 *If M' is a submodule of M then $\gamma(\mathrm{Gr}(M')) \le \gamma(\mathrm{Gr}(M))$.*

Proof We compare $\mathrm{Gr}(M')$ with the graded module $\mathrm{Gr}_M(M')$ given by

$$\mathrm{Gr}_M(M') = \bigoplus_{j=0}^{\infty} (M' \cap \mathfrak{M}^j M)/(M' \cap \mathfrak{M}^{j+1}M).$$

Since $\mathrm{Gr}(A)$ is Noetherian, $\mathrm{Gr}_M(M')$ is a finitely generated $\mathrm{Gr}(A)$-module. If it is generated by elements of degree at most m, then for $j \geq 0$ we have

$$M' \cap \mathfrak{M}^{m+j}M = \mathfrak{M}^j(M' \cap \mathfrak{M}^m M)$$

and so

$$M' \cap \mathfrak{M}^{m+j}M \subseteq \mathfrak{M}^j M' \subseteq M' \cap \mathfrak{M}^j M.$$

It follows that

$$\gamma(\mathrm{Gr}(M')) = \gamma(\mathrm{Gr}_M(M')) \leq \gamma(\mathrm{Gr}(M)).$$

□

Theorem 2.3.2 *Suppose that A is a commutative Noetherian local ring and that M is a finitely generated A-module. Then the following are equal:*
(i) *The Krull dimension of M as an A-module,*
(ii) *The Krull dimension of* $\mathrm{Gr}(M)$ *as a* $\mathrm{Gr}(A)$*-module,*
(iii) $\gamma(\mathrm{Gr}(M))$,
(iv) *The length n of any system $x_1, \dots, x_n$ of parameters for M as an A-module.*

Proof By Theorem 2.2.7, $\gamma(\mathrm{Gr}(M))$ is equal to the Krull dimension of $\mathrm{Gr}(M)$ as a $\mathrm{Gr}(A)$-module. We have already seen above that

$$\gamma(\mathrm{Gr}(M)) \leq n \leq \dim(M),$$

and so it remains to prove that $\dim(M) \leq \gamma(\mathrm{Gr}(M))$. We prove this by induction on $\gamma(\mathrm{Gr}(M))$. If this is zero, then by Proposition 2.2.3 we have $\dim(M) = 0$, so we may assume that $\gamma(\mathrm{Gr}(M)) \geq 1$. By Proposition 2.2.5 (ii), we can choose an associated prime $\mathfrak{p}$ of M with $\dim(A/\mathfrak{p}) = \dim(M)$. Since $A/\mathfrak{p}$ is isomorphic to a submodule of M, by the lemma we have $\gamma(\mathrm{Gr}(A/\mathfrak{p})) \leq \gamma(\mathrm{Gr}(M))$, so it suffices to deal with the case $M = A/\mathfrak{p}$. Let $\mathfrak{p} = \mathfrak{p}_0 \supset \mathfrak{p}_1 \supset \cdots \supset \mathfrak{p}_r$ be a chain of prime ideals in A with $r = \dim(A/\mathfrak{p})$. Choose $x \in \mathfrak{p}_{r-1}$ with $x \notin \mathfrak{p}_r$. Then $\dim(M/xM) = \dim(M) - 1$ while $\gamma(\mathrm{Gr}(M/xM)) \leq \gamma(\mathrm{Gr}(M)) - 1$, and so this completes the proof by induction. □

Remark It follows from this theorem that a sequence $x_1, \dots, x_n$ of elements of A is a system of parameters for M if and only if M has dimension n and $M/(x_1, \dots, x_n)M$ has dimension zero. Thus our definition agrees with the one given in Matsumura [59].

The **height** of a prime ideal $\mathfrak{p}$ is the maximum length n of a chain of proper inclusions of prime ideals $\mathfrak{p} = \mathfrak{p}_0 \supset \mathfrak{p}_1 \cdots \supset \mathfrak{p}_n$, or ∞ if there are such chains of unbounded length. Thus in an integral domain the prime ideals of height one are the minimal non-zero primes.

Corollary 2.3.3 (Krull's principal ideal theorem) *Suppose that A is a commutative Noetherian ring, and $(x) \neq 0$ is a principal ideal in A. Then any prime ideal $\mathfrak{p}$ which is minimal among the primes containing (x) has height at most one.*

Proof Let x_1 be the image of x in the localization $A_\mathfrak{p}$. Since the prime ideals in $A_\mathfrak{p}$ correspond to the prime ideals in A contained in $\mathfrak{p}$, it suffices to prove this for the ideal (x_1) of $A_\mathfrak{p}$. Now $A_\mathfrak{p}/(x_1)$ has Krull dimension zero (i.e., finite length), and so $A_\mathfrak{p}$ has a system of parameters of length at most one. So by the theorem $A_\mathfrak{p}$ has Krull dimension at most one, as required. □

2.4 Degree and ψ

We saw in Section 2.2 that if $A = \bigoplus_{j=0}^{\infty} A_j$ is a finitely generated commutative graded K-algebra with $A_0 = K$, then the Krull dimension of A is equal to the order of the pole of the rational function $p(A,t)$ at $t = 1$. If the Krull dimension is n, then the value of the rational function $(1-t)^n p(A,t)$ at $t = 1$ is a non-zero rational number, called the **degree** of A, written $\deg(A)$. More generally, if $M = \bigoplus_{j=-\infty}^{\infty} M_j$ is a finitely generated graded A-module, we define rational numbers $\deg(M)$ and $\psi(M)$ by the Laurent expansion about $t = 1$:

$$p(M,t) = \frac{\deg(M)}{(1-t)^n} + \frac{\psi(M)}{(1-t)^{n-1}} + O\left(\frac{1}{(1-t)^{n-2}}\right).$$

Clearly deg and ψ are additive over short exact sequences of modules.

Now suppose that A is an integral domain, with field of fractions L. We define the **rank** of M to be the dimension over L of $L \otimes_A M$. We write $M[d]$ for M with a degree shift of d, so that $M[d]_i = M_{i+d}$.

Lemma 2.4.1 (i) $\deg(M[d]) = \deg(M)$.

(ii) *If M has Krull dimension at most $n-1$ then* $\deg(M) = 0$.

(iii) $\deg(M) = \mathrm{rank}_A(M)\deg(A)$.

(iv) $\psi(M[d]) = \psi(M) - d\deg(M)$.

(v) *If M has Krull dimension at most $n-2$ then* $\psi(M) = 0$.

(vi) *If* $\deg(M) = 0$ *then* $\psi(M) = \sum_\mathfrak{p} \mathrm{length}_{A_\mathfrak{p}}(M_\mathfrak{p})\psi(A/\mathfrak{p})$ *where the sum runs over the homogeneous height one primes $\mathfrak{p}$ of A.*

Proof Parts (i) and (iv) follow from the formula $p(M[d],t) = t^{-d}p(M,t)$. Parts (ii) and (v) are clear. To prove parts (iii) and (vi), note that both sides are additive over short exact sequences of graded modules, so it suffices to consider modules of the form $(A/\mathfrak{p})[d]$ with $\mathfrak{p}$ a homogeneous prime in A. □

Proposition 2.4.2 *Suppose that $B \supseteq A$ is a finite extension of graded integral domains finitely generated over K by elements of positive degree, and that $L' \supseteq L$ are their fields of fractions. Then*

$$\deg(B) = |L' : L| \deg(A).$$

Proof This follows from part (iii) of the lemma. □

Theorem 2.4.3 *Suppose that G is a finite group and V is an n-dimensional faithful representation of G over a field K. Then $p(K[V]^G, t)$ is a rational function of t with a pole at $t = 1$ of order n. The value at $t = 1$ of the rational function $(1-t)^n p(K[V]^G, t)$ is equal to $1/|G|$.*

Proof We saw in Chapter 1 that $K[V]^G$ is finitely generated over K, so that rationality follows from Theorem 2.1.1, and the order of the pole at $t = 1$ was computed in Proposition 2.1.4. Now $K[V]$ is a finite extension of $K[V]^G$, so we may apply Proposition 2.4.2. At the level of fields of fractions, $K(V)$ is a Galois extension of $K(V)^G$ with Galois group G. So we conclude from the proposition that

$$\deg K[V] = |G|. \deg K[V]^G.$$

Since $p(K[V], t) = 1/(1-t)^n$, $\deg K[V] = 1$, and the theorem is proved. □

2.5 Molien's theorem

Molien's theorem is a formula for the Poincaré series of the invariants in characteristic zero.

Suppose that K is a field of characteristic zero. Then the map $\pi_G = \frac{1}{|G|}\sum_{g\in G} g$ is a projection on $K[V]$ with image $K[V]^G$. So the dimension of $K[V]^G_j$ is equal to the trace of the matrix representing π on $K[V]_j$. So we have

$$p(K[V]^G, t) = \frac{1}{|G|}\sum_{g\in G}\sum_{j=0}^{\infty} \mathrm{Tr}(g, K[V]_j)t^j.$$

The following formula for the trace actually works independently of the characteristic of K.

Proposition 2.5.1 $\displaystyle\sum_{j=0}^{\infty} t^j \mathrm{Tr}(g, K[V]_j) = \frac{1}{\det(1 - g^{-1}t, V)}.$

To make sense of this, note that $\mathrm{Mat}_n(K)[t] \cong \mathrm{Mat}_n(K[t])$, so that det makes sense on $\mathrm{Mat}_n(K)[t]$.

Proof Extending the field does not affect either side of this equation, so we may assume that K is algebraically closed. Then g is upper triangularizable on V (if K has characteristic zero, it is even diagonalizable), say with eigenvalues $\lambda_1, \dots, \lambda_n$. The eigenvalues on V^* are $\lambda_1^{-1}, \dots, \lambda_n^{-1}$, so the eigenvalues on $K[V]_j$ are the products of j not necessarily distinct λ_i^{-1}'s. So we have

$$\sum_{j=0}^{\infty} t^j \mathrm{Tr}(g, K[V]_j) = (\sum_{j=0}^{\infty} \lambda_1^{-j} t^j) \cdots (\sum_{j=0}^{\infty} \lambda_n^{-j} t^j) = \prod_{i=1}^{n} \frac{1}{1-\lambda_i^{-1}t} = \frac{1}{\det(1-g^{-1}t, V)},$$

which proves the proposition. □

Example If $g = 1$ then this says that $\sum_{j=0}^{\infty} t^j \dim_K K[V]_j = \frac{1}{(1-t)^n}$.

Theorem 2.5.2 (Molien) *If K is a field of characteristic zero, then we have*

$$p(K[V]^G, t) = \frac{1}{|G|} \sum_{g \in G} \frac{1}{\det(1-g^{-1}t, V)}.$$

Proof This follows immediately from the proposition and the formula immediately preceding it. □

Remark The left hand side of the equation given by Molien's theorem lies in $\mathbb{Q}(t)$, while the right hand side lies in $K(t)$, so the equation only makes sense if K has characteristic zero. However, in non-zero characteristic coprime to $|G|$, Molien's theorem still holds, as long as one uses a "Brauer lift" of the trace and determinant, obtained by lifting eigenvalues to characteristic zero and adding or multiplying. Even in non-coprime characteristic, Molien's formula gives information. It no longer calculates the Poincaré series of the invariants, but rather the Poincaré series for the multiplicity of the trivial module as a composition factor.

The corresponding formula for relative invariants is as follows.

Theorem 2.5.3 *If K is a field of characteristic zero and S is a simple KG-module, then*

$$p(K[V]^G_S, t) = \frac{\dim_K(S)}{|G|} \sum_{g \in G} \frac{\mathrm{Tr}(g^{-1}, S)}{\det(1-g^{-1}t, V)}.$$

Proof By character theory, we have

$$p(K[V]^G_S, t) = \sum_{j=0}^{\infty} t^j \dim_K(K[V]^G_S)_j$$

$$= \sum_{j=0}^{\infty}\left(t^j \frac{\dim_K(S)}{|G|}\sum_{g\in G}\mathrm{Tr}(g^{-1},S)\mathrm{Tr}(g,K[V]_j)\right)$$

$$= \frac{\dim_K(S)}{|G|}\sum_{g\in G}\left(\mathrm{Tr}(g^{-1},S)\sum_{j=0}^{\infty}t^j\mathrm{Tr}(g,K[V]_j)\right)$$

$$= \frac{\dim_K(S)}{|G|}\sum_{g\in G}\frac{\mathrm{Tr}(g^{-1},S)}{\det(1-g^{-1}t,V)}.$$

In the last step, we have used Proposition 2.5.1. □

Exercise Let G be the quaternion group Q_8, and let V be the irreducible two dimensional representation of G over the complex numbers. Find the eigenvalues of the elements of G on V, and use Molien's formula to prove that

$$p(\mathbb{C}[V]^G,t) = \frac{1+t^6}{(1-t^4)^2} = 1+2t^4+t^6+\cdots$$

Find two invariants of degree four and an invariant of degree six. Examine the ring they generate, and by comparing Poincaré series, deduce that

$$\mathbb{C}[V]^G = \mathbb{C}[\alpha,\beta,\gamma]/(\gamma^2-\alpha^2\beta+4\beta^3),$$

with α and β elements of degree four and γ an element of degree six.

Show that in this example, the Poincaré series for the relative invariants corresponding to the simple module V is given by $p(\mathbb{C}[V]^G_V,t) = t/(1-t^2)^2$.

2.6 Reflecting hyperplanes

We saw in Section 2.4 that if V is a faithful representation of G with $\dim_K(V)=n$, then the value at $t=1$ of the rational function $(1-t)^n p(K[V]^G,t)$ is equal to $1/|G|$. In characteristic zero, one may also deduce this from Molien's theorem. Namely, if $g\in G$ has eigenvalues $\lambda_1,\dots,\lambda_n$, then

$$\frac{1}{\det(1-g^{-1}t,V)} = \prod_{i=1}^{n}\frac{1}{1-\lambda_i^{-1}t},$$

so that the order of the pole at $t=1$ is equal to the number of λ_i which are equal to one, namely the dimension of the fixed space V^g. So if $g\neq 1$, then after multiplying by $(1-t)^n$, we have a zero at $t=1$ (of order $n-\dim_K(V^g)$). So when we evaluate

$$(1-t)^n p(K[V]^G,t) = \frac{1}{|G|}\sum_{g\in G}\frac{(1-t)^n}{\det(1-g^{-1}t,V)}$$

at $t = 1$, the only contribution comes from the identity element. This contribution is $1/|G|$, as required.

We can obtain $\psi(K[V]^G)$ (see Section 2.4) by the same method. Namely, once the contribution from the identity element has been subtracted out, the remaining expression has a pole of order at most $n-1$ at $t=1$. An element $g \in G$ contributes to this pole if and only if $\dim_K(V^g) = n-1$. An element $1 \neq g \in G$ with this property is called a **pseudoreflection,** and its fixed space is called a **reflecting hyperplane.** So a pseudoreflection has eigenvalue 1 with multiplicity $n-1$, and one other eigenvalue, say λ. So we have

$$\frac{1}{\det(1-g^{-1}t, V)} = \frac{1}{(1-t)^{n-1}(1-\lambda^{-1}t)}$$

and after multiplying by $(1-t)^{n-1}$, the value at $t=1$ is $1/(1-\lambda^{-1})$. Next, we observe that

$$\frac{1}{1-\lambda^{-1}} + \frac{1}{1-\lambda} = \frac{(1-\lambda)+(1-\lambda^{-1})}{1-\lambda^{-1}-\lambda+1} = 1.$$

So if we pair together each element with its inverse, we see that the total contribution from all pseudoreflections is equal to one half of the number of them. Self-inverse elements have $\lambda = -1$, so that their contribution is also one half. Thus the Laurent expansion begins

$$p(K[V]^G, t) = \frac{1}{|G|}\left(\frac{1}{(1-t)^n} + \frac{r/2}{(1-t)^{n-1}} + \cdots\right)$$

where r is the number of pseudoreflections in G. So

$$\psi(K[V]^G) = \frac{r}{2|G|}.$$

Carlisle and Kropholler [21] have made a conjecture which gives an analogous formula in case $K = \mathbb{F}_p$. Namely, suppose that V is a faithful representation of G. For each subspace $W \subset V$ of codimension one, we let G_W denote the pointwise stabilizer of W in G. We set $|G_W| = p^{a_W}.h_W$ with h_W coprime to p. The Carlisle–Kropholler conjecture states that

$$\psi(K[V]^G) = \sum_W ((p-1)a_W + h_W - 1)$$

where the sum runs over all hyperplanes W for which $|G_W| > 1$. This conjecture was proved recently by Benson and Crawley-Boevey [14]. We shall present this proof in Section 3.13.

Chapter 3

Divisor Classes, Ramification and Hyperplanes

In this chapter, we calculate the divisor class group $\mathrm{Cl}(K[V]^G)$. The approach uses Samuel's theory of Galois descent [86]. The final theorem, proved in Section 3.9, states that $\mathrm{Cl}(K[V]^G)$ is isomorphic to the subgroup of $\mathrm{Hom}(G, K^\times)$ consisting of those homomorphisms which take the value one on every pseudoreflection. As a consequence, we prove a theorem of Nakajima, which states that $K[V]^G$ is a unique factorization domain if and only if this subgroup of $\mathrm{Hom}(G, K^\times)$ is trivial. In contrast, the Picard group $\mathrm{Pic}(K[V]^G)$ is always trivial. We prove this theorem of Kang in Section 3.6.

We also give a ramification formula for the invariant ψ introduced in Section 2.4, and use this to prove the Carlisle–Kropholler conjecture in Section 3.13.

We begin with some generalities on divisors. Our treatment follows Samuel [86], with the exception that there are some minor simplifications arising from the fact that we are not interested in non-Noetherian rings here.

3.1 Divisors

We say that a ring A is a **normal domain** if it is a commutative Noetherian integrally closed domain (these hypotheses are satisfied by $K[V]$ and $K[V]^G$ by Proposition 1.1.1 and Theorem 1.3.1). Let A be a normal domain, and let L be the field of fractions of A. A **fractional ideal** $\mathfrak{a}$ of A is a non-zero A-submodule of L with the property that there exists a non-zero element $x \in A$ with $x\mathfrak{a} \subseteq A$. Since A is Noetherian, this is equivalent to the statement that $\mathfrak{a}$ is finitely generated as an A-module. We say that $\mathfrak{a}$ is **principal** if $\mathfrak{a} = (x)$ is generated by a single element $x \in L$ as an A-module. We say that $\mathfrak{a}$ is **divisorial** if it is an intersection of principal fractional ideals.

For a fractional ideal $\mathfrak{a}$, we write

$$\mathfrak{a}^{-1} = \{b \in L \mid b\mathfrak{a} \subseteq A\} = \bigcap_{a \in \mathfrak{a}} (a^{-1}).$$

By definition, this is a divisorial fractional ideal. So $(\mathfrak{a}^{-1})^{-1}$ is again a divisorial fractional ideal, containing $\mathfrak{a}$. Furthermore, $(\mathfrak{a}^{-1})^{-1}$ is contained in every principal fractional ideal containing $\mathfrak{a}$, and so it is the smallest divisorial fractional ideal containing $\mathfrak{a}$. So we write $\bar{\mathfrak{a}}$ for $(\mathfrak{a}^{-1})^{-1}$ and regard $\bar{\mathfrak{a}}$ as the "divisorialization" of $\mathfrak{a}$.

Proposition 3.1.1 *If A is a normal domain and $\mathfrak{a}$ is a divisorial fractional ideal, then $\overline{(\mathfrak{a}\mathfrak{a}^{-1})} = A$.*

Proof It is clear that $\mathfrak{a}\mathfrak{a}^{-1} \subseteq A$, so we must show that if $\mathfrak{a}$ is divisorial and $\mathfrak{a}\mathfrak{a}^{-1} \subseteq (x)$ then $A \subseteq (x)$. Now $\mathfrak{a}$ is contained in some principal fractional ideal (y), so by replacing $\mathfrak{a}$ by $y^{-1}\mathfrak{a}$ and $\mathfrak{a}^{-1}$ by $y\mathfrak{a}^{-1}$, we may assume that $\mathfrak{a} \subseteq A$.

If $\mathfrak{a}\mathfrak{a}^{-1} \subseteq (x)$ then $x^{-1}\mathfrak{a} \subseteq (\mathfrak{a}^{-1})^{-1} = \mathfrak{a}$ and so by induction, $x^{-n}\mathfrak{a} \subseteq \mathfrak{a}$ for all $n > 0$. Choose a non-zero element $a \in \mathfrak{a}$. Since A is Noetherian, the ascending chain of ideals $(a, x^{-1}a, \dots, x^{-n}a)$ $(n > 0)$ must terminate, so for some $n > 0$ we have $x^{-n}a \in (a, x^{-1}a, \dots, x^{-n+1}a)$. Dividing by a, we see that x^{-1} satisfies some monic polynomial with coefficients in A. Since A is integrally closed, this means that $x^{-1} \in A$, so that $A \subseteq (x)$ as required. □

This allows us to define the group of divisors as follows. We say that two fractional ideals $\mathfrak{a}$ and $\mathfrak{b}$ are **Artin equivalent** (written $\mathfrak{a} \sim \mathfrak{b}$) if $\bar{\mathfrak{a}} = \bar{\mathfrak{b}}$, or equivalently $\mathfrak{a}^{-1} = \mathfrak{b}^{-1}$. A **divisor** is an Artin equivalence class of fractional ideals. We write $d(\mathfrak{a})$ for the divisor determined by $\mathfrak{a}$.

Lemma 3.1.2 *If $\mathfrak{a} \sim \mathfrak{b}$ then $\mathfrak{a}\mathfrak{c} \sim \mathfrak{b}\mathfrak{c}$.*

Proof If $\mathfrak{a} \sim \mathfrak{b}$ then $\mathfrak{a}^{-1} = \mathfrak{b}^{-1}$. Now observe that

$$(\mathfrak{a}\mathfrak{c})^{-1} = \{x \in L \mid x\mathfrak{c} \subseteq \mathfrak{a}^{-1}\} = \{x \in L \mid x\mathfrak{c} \subseteq \mathfrak{b}^{-1}\} = (\mathfrak{b}\mathfrak{c})^{-1}$$

so that $\mathfrak{a}\mathfrak{c} \sim \mathfrak{b}\mathfrak{c}$. □

It follows that multiplication of fractional ideals passes down to a well defined composition on divisors, which we write additively: $d(\mathfrak{a}) + d(\mathfrak{b}) = d(\mathfrak{a}\mathfrak{b})$. By Proposition 3.1.1, with this addition, the divisors form an abelian group with $d(A)$ as the zero element and $d(\mathfrak{a}^{-1}) = -d(\mathfrak{a})$. This is the **divisor group** $D(A)$. Now if $\mathfrak{a} \subseteq \mathfrak{b}$ then $\bar{\mathfrak{a}} \subseteq \bar{\mathfrak{b}}$, and so inclusion passes down to a well defined partial order $\geq$ (note the reversal) on divisors, which is compatible with addition in the sense that $d(\mathfrak{a}) \geq d(\mathfrak{b})$ implies $d(\mathfrak{a}) + d(\mathfrak{c}) \geq d(\mathfrak{b}) + d(\mathfrak{c})$. Thus $D(A)$ is an **ordered abelian group.**

Since A is Noetherian, the positive divisors satisfy the descending chain condition with respect to $\geq$. Also, since $d(\mathfrak{a} \cap \mathfrak{b}) \geq d(\mathfrak{a})$, any two elements have a least upper bound.

Lemma 3.1.3 *In an ordered abelian group with least upper bounds $lub(a,b)$, greater lower bounds $glb(a,b)$ also exist, and $glb(a,b) = a + b - lub(a,b)$.*

If c is a minimal strictly positive element and a and b are positive elements with $c \leq a + b$ then either $c \leq a$ or $c \leq b$.

Proof Clearly $a + b - lub(a,b)$ is a lower bound for a and b. Conversely, if $c \leq a$ and $c \leq b$ then $a \leq a + b - c$ and $b \leq a + b - c$, and so $lub(a,b) \leq a + b - c$. Thus $c \leq a + b - lub(a,b)$.

If c is a minimal strictly positive element and $c \leq a+b$ but $c \not\leq a$, then $glb(a,c) = 0$, and so $lub(a,c) = a + c$. Since $a \leq a + b$ and $c \leq a + b$ it follows that $a + c = lub(a,c) \leq a + b$ and hence $c \leq b$. □

Proposition 3.1.4 *Suppose that D is an ordered abelian group satisfying the descending chain condition on positive elements, and with the property that every pair of elements has a least upper bound and a greatest lower bound. Then D is isomorphic, as an ordered abelian group, to the free abelian group on the minimal strictly positive elements.*

Proof Since every pair of elements has a least upper bound, given $x \in D$, let $z = lub(x,0)$. Then $x = z-(z-x)$ is a difference of positive elements, so D is generated by the positive elements. Since D satisfies the descending chain condition, this means that D is generated by the minimal strictly positive elements. If x_i are strictly positive elements and $\sum_{i\in I} n_i x_i \geq 0$ with $n_i \in \mathbb{Z}$ and I a finite indexing set, we shall show that each $n_i \geq 0$, so that in particular if $\sum_{i\in I} n_i x_i = 0$ then each $n_i = 0$. Let $I' = \{i \in I \mid n_i > 0\}$, $I'' = \{i \in I \mid n_i < 0\}$. Then $\sum_{i\in I'} n_i x_i \geq \sum_{i \in I''}(-n_i)x_i$. If $I'' \neq \emptyset$ then for $j \in I''$ we have $x_j \leq \sum_{i\in I'} n_i x_i$, so that applying the above lemma, we have $x_j \leq x_i$ for some $i \in I'$. This contradicts the minimality of x_i, and so $I'' = \emptyset$ and the x_i freely generate D. □

The divisors in $D(A)$ which are minimal strictly positive with respect to $\geq$ are called the **prime divisors.**

Corollary 3.1.5 *If A is a normal domain then $D(A)$ is the free abelian group on the prime divisors.* □

3.2 Primes of height one

Next, we examine the prime divisors in more detail, and show that they correspond to the prime ideals of height one in A.

Lemma 3.2.1 *If $\mathfrak{p}$ is a divisorial ideal then $d(\mathfrak{p})$ is a prime divisor if and only if $\mathfrak{p}$ is a prime ideal in A.*

Proof Suppose that $d(\mathfrak{p})$ is a prime divisor. Since $d(\mathfrak{p}) \geq 0$, $\mathfrak{p} \subseteq A$. If $x, y \in A$ and $xy \in \mathfrak{p}$, then $d(x) + d(y) = d(xy) \geq d(\mathfrak{p})$ and so by Lemma 3.1.3, either $d(x) \geq d(\mathfrak{p})$ or $d(y) \geq d(\mathfrak{p})$. So either $x \in \mathfrak{p}$ or $y \in \mathfrak{p}$.

Conversely, if $\mathfrak{p}$ is a prime ideal in A, write $d(\mathfrak{p}) = \sum n_i \mathfrak{p}_i$ with $d(\mathfrak{p}_i)$ prime divisors, so that the $\mathfrak{p}_i$ are prime ideals. Then $\mathfrak{p} = \prod \mathfrak{p}_i^{n_i}$, so that for some i, $\mathfrak{p} = \mathfrak{p}_i$. □

Lemma 3.2.2 *Every non-zero prime ideal contains a non-zero prime ideal which is divisorial. There are no strict containments between divisorial ideals.*

Proof Let $\mathfrak{p} \neq 0$ be a prime ideal, and choose $0 \neq x \in \mathfrak{p}$. By Proposition 3.1.4 we may write $0 \neq d(x) = \sum_i n_i d(\mathfrak{p}_i)$ with $n_i \geq 0$ and $\mathfrak{p}_i$ divisorial ideals with $d(\mathfrak{p}_i)$ prime divisors. By Lemma 3.2.1, the $\mathfrak{p}_i$ are prime ideals. If $y \in \prod \mathfrak{p}_i^{n_i}$ then $d(y) \geq d(x)$ and so $(y) \subseteq (x)$. Thus $\prod \mathfrak{p}_i^{n_i} \subseteq (x) \subseteq \mathfrak{p}$. Since $\mathfrak{p}$ is prime, this means that some $\mathfrak{p}_i \subseteq \mathfrak{p}$.

It now follows from Lemma 3.2.1 that there are no strict containments between divisorial prime ideals. □

Proposition 3.2.3 *A prime ideal is divisorial if and only if it has height one. The prime divisors are the divisors of the form $d(\mathfrak{p})$, where $\mathfrak{p}$ is a prime ideal of height one.*

Proof If $\mathfrak{p}$ has height one, then by the above lemmas $\mathfrak{p}$ is divisorial and $d(\mathfrak{p})$ is minimal strictly positive in $D(A)$. Conversely, if $\mathfrak{p}$ properly contains a non-zero prime ideal then it properly contains one which is divisorial, which is impossible. □

Theorem 3.2.4 *$D(A)$ is the free abelian group on the set of divisors $d(\mathfrak{p})$, where $\mathfrak{p}$ runs over the prime ideals in A of height one.*

Proof This follows from Corollary 3.1.5 and Proposition 3.2.3. □

Example **A Dedekind domain** A is a normal domain of Krull dimension one. In such a ring, every non-zero prime ideal is maximal and has height one. So $D(A)$ is the free abelian group on the non-zero prime ideals.

Theorem 3.2.5 *If A is a Dedekind domain then every non-zero ideal $\mathfrak{a}$ is divisorial, and satisfies $\mathfrak{a}\mathfrak{a}^{-1} = A$. If, furthermore, A is local with maximal ideal $\mathfrak{p}$, then every non-zero ideal is a power of $\mathfrak{p}$.*

Proof By Lemma 3.2.2, every non-zero prime ideal $\mathfrak{p}$ is divisorial, and by maximality it satisfies $\mathfrak{p}\mathfrak{p}^{-1} = A$. If there is a non-zero ideal which does not satisfy both of these properties, let $\mathfrak{a}$ be maximal among ideals like this. Then $\mathfrak{a}$ is properly contained in some non-zero prime ideal $\mathfrak{p}$. We claim that $\mathfrak{a}\mathfrak{p}^{-1}$ properly contains $\mathfrak{a}$. If

$\mathfrak{a}\mathfrak{p}^{-1} = \mathfrak{a}$ then $\mathfrak{a} = \mathfrak{a}\mathfrak{p}^{-1}\mathfrak{p} = \mathfrak{a}\mathfrak{p}$. Since $\mathfrak{a} = (a_1, \ldots, a_n)$ is finitely generated, we then have $a_i = \sum \lambda_{ij} a_j$ with $\lambda_{ij} \in \mathfrak{p}$. Multiplying by the transposed matrix of cofactors of the matrix $I - (\lambda_{ij})$, we see that $\det(I - (\lambda_{ij}))$ annihilates the a_i, so that since A is an integral domain it is zero. This contradicts the fact that this determinant is congruent to one modulo $\mathfrak{p}$, and so $\mathfrak{a}\mathfrak{p}^{-1}$ properly contains $\mathfrak{a}$ as claimed. By choice of $\mathfrak{a}$, it follows that $\mathfrak{a}\mathfrak{p}^{-1}$ is divisorial. So $\mathfrak{a}\mathfrak{p}^{-1}(\mathfrak{a}\mathfrak{p}^{-1})^{-1}$ is also divisorial, so that by Proposition 3.1.1 we have $\mathfrak{a}\mathfrak{p}^{-1}(\mathfrak{a}\mathfrak{p}^{-1})^{-1} = \mathfrak{a}$. Thus

$$\mathfrak{a}^{-1} \supseteq \mathfrak{p}^{-1}(\mathfrak{a}\mathfrak{p}^{-1})^{-1} \supseteq \mathfrak{p}^{-1}\mathfrak{a}^{-1}(\mathfrak{p}^{-1})^{-1} = \mathfrak{p}^{-1}\mathfrak{a}\mathfrak{p} = \mathfrak{a}^{-1}$$

and so $\mathfrak{a}^{-1} = \mathfrak{p}^{-1}(\mathfrak{a}\mathfrak{p}^{-1})^{-1}$, which implies that $\mathfrak{a}\mathfrak{a}^{-1} = A$. Thus $(\mathfrak{a}^{-1})^{-1} = (\mathfrak{a}^{-1})^{-1}\mathfrak{a}^{-1}\mathfrak{a} \subseteq (\mathfrak{a}^{-1}\mathfrak{a})^{-1}\mathfrak{a} = \mathfrak{a}$ and so $\mathfrak{a}$ is divisorial. Finally if A is local with maximal ideal $\mathfrak{p}$, then $D(A) \cong \mathbf{Z}$ and the last statement of the theorem follows. □

Notation If A is a normal domain and $0 \neq x \in L$, then the principal fractional ideal (x) is divisorial, and so by Theorem 3.2.4, there exist uniquely defined integers $v_\mathfrak{p}(x) \in \mathbf{Z}$ such that

$$d(x) = \sum_\mathfrak{p} v_\mathfrak{p}(x) d(\mathfrak{p}).$$

Here, $\mathfrak{p}$ runs over the prime ideals of height one, and $v_\mathfrak{p}(x)$ is zero for all but a finite number of $\mathfrak{p}$, so that this sum is finite. We also set $v_\mathfrak{p}(0) = \infty$.

The function $v_\mathfrak{p} : L \to \mathbf{Z} \cup \{\infty\}$ is called the **discrete valuation** associated with $\mathfrak{p}$. Since $d(xy) = d(x) + d(y)$ and $d(x+y) \geq glb(d(x), d(y))$ we have

(i) $v_\mathfrak{p}(xy) = v_\mathfrak{p}(x) + v_\mathfrak{p}(y)$

(ii) $v_\mathfrak{p}(x+y) \geq glb(v_\mathfrak{p}(x), v_\mathfrak{p}(y))$

(iii) $v_\mathfrak{p}(x) = \infty \Leftrightarrow x = 0$.

Conversely, if $v : L \to \mathbf{Z} \cup \{\infty\}$ is any function satisfying (i), (ii) and (iii), then $\{x \in A \mid v(x) > 0\}$ is a prime ideal, $\mathfrak{p}$ say. If $\mathfrak{p}$ has height one then the localization $A_\mathfrak{p}$ is a maximal subring of the field L, and so $\{x \in L \mid v(x) \geq 0\} = A_\mathfrak{p}$. Now $A_\mathfrak{p}$ is a local Dedekind domain, which by Theorem 3.2.5 implies that the only ideals are the powers of $\mathfrak{p}_\mathfrak{p}$, and so v must be some positive integer multiple of $v_\mathfrak{p}$.

Corollary 3.2.6 *Suppose that A is a normal domain. If $\mathfrak{a}$ is a fractional ideal, then $\bar{\mathfrak{a}} = \bigcap_\mathfrak{p} \mathfrak{a}_\mathfrak{p}$, where $\mathfrak{p}$ runs over the prime ideals in A of height one, and the intersection takes place in the field of fractions L of A. In particular, we have $A = \bigcap_\mathfrak{p} A_\mathfrak{p}$, with $\mathfrak{p}$ as before.*

Proof Write $d(\mathfrak{a}) = \sum_\mathfrak{p} n_\mathfrak{p} d(\mathfrak{p})$, where $\mathfrak{p}$ runs over the prime ideals of height one and $n_\mathfrak{p}$ is zero for all but a finite number of such $\mathfrak{p}$. Then given $x \in L$, since (x) is divisorial, we have

$$x \in \bar{\mathfrak{a}} \Leftrightarrow (x) \subseteq \bar{\mathfrak{a}} \Leftrightarrow d(x) \geq d(\mathfrak{a}) \Leftrightarrow v_\mathfrak{p}(x) \geq n_\mathfrak{p} \ \forall \mathfrak{p} \Leftrightarrow x \in \bigcap_\mathfrak{p} \mathfrak{a}_\mathfrak{p}.$$

□

3.3 Duality

If M and N are modules for a commutative ring A, we write $\mathrm{Hom}_A(M,N)$ for the A-module homomorphisms. This may be regarded as an A-module, with action given by $(af)(m) = a.f(m)$ for $a \in A$, $m \in M$ and $f \in \mathrm{Hom}_A(M,N)$. We write M^* for the dual module

$$M^* = \mathrm{Hom}_A(M,A).$$

If A, M and N are graded, we may also form the graded dual, whose degree i part consists of the homomorphisms which raise degree by i.

Lemma 3.3.1 *Suppose that $A = \bigoplus_{j=0}^{\infty} A_j$ is a commutative graded ring with $A_0 = K$, and finitely generated by elements of positive degree. If M and N are finitely generated graded A-modules then the inclusion of the graded homomorphisms from M to N in the ungraded homomorphisms is an isomorphism.*

Proof This is true when M is freely generated over A by a finite set of homogeneous elements, since $\mathrm{Hom}_A(A,N) = N$. So choose an epimorphism $F \to M$ of graded modules, with F freely generated over A by a finite set of homogeneous elements. Then the ungraded homomorphisms from M to N are contained in the graded homomorphisms $\mathrm{Hom}_A(F,N)$. If $\phi = \sum_i \phi_i$ is an ungraded homomorphism from M to N with $\phi_i \in \mathrm{Hom}_A(F,N)_i$, then each ϕ_i vanishes on the kernel of $F \to M$, and so each ϕ_i is in $\mathrm{Hom}_A(M,N)$. □

It follows from this lemma that the modules $\mathrm{Ext}^r_A(M,N)$ may also be regarded as graded A-modules. We have the following duality formula [14] for the invariant ψ introduced in Section 2.4:

Theorem 3.3.2 *Suppose that $A = \bigoplus_{j=0}^{\infty} A_j$ is a graded normal domain of Krull dimension n, with $A_0 = K$. If M and N are finitely generated graded A-modules then*

$$\psi(\mathrm{Hom}_A(M,N)) - \psi(\mathrm{Ext}^1_A(M,N)) =$$
$$\mathrm{rank}_A(M)\mathrm{rank}_A(N)\psi(A) + \mathrm{rank}_A(M)\psi(N) - \mathrm{rank}_A(N)\psi(M).$$

Proof The right hand side of the equation is clearly additive on short exact sequences $0 \to M \to M' \to M'' \to 0$ of graded modules. To see that the same is true of the left hand side, we argue as follows. Since A is integrally closed, the localization at any height one prime is a Dedekind domain, and therefore has the property that any submodule of a projective module is projective. Thus $\mathrm{Ext}^2_A(M'',N)$ has Krull dimension at most $n-2$, and hence so does the image of the last map in the exact sequence

$$0 \to \mathrm{Hom}_A(M'',N) \to \mathrm{Hom}_A(M',N) \to \mathrm{Hom}_A(M,N) \to$$

$$\mathrm{Ext}^1_A(M'',N) \to \mathrm{Ext}^1_A(M',A) \to \mathrm{Ext}^1_A(M,N) \to \mathrm{Ext}^2_A(M'',N).$$

It follows that ψ vanishes on the image of this last map, and so $\psi(\mathrm{Hom}_A(M,N)) - \psi(\mathrm{Ext}^1_A(M,N))$ is additive on short exact sequences as required.

In the same way, one sees that the left and right hand side of the equation are additive on short exact sequences $0 \to N \to N' \to N'' \to 0$ in the second variable. It follows that it suffices to prove the theorem with M and N of the form $(A/\mathfrak{p})[d]$ with $\mathfrak{p}$ a homogeneous prime ideal in A. We divide into cases according to the forms of the primes involved. First, we note that if either prime has height two or more, all terms in the equation are zero.

In the case where $M = A[d]$ we have $\mathrm{Hom}_A(M,N) \cong N[-d]$ and $\mathrm{Ext}^1_A(M,N) = 0$, and so by Lemma 2.4.1, both sides of the equation are equal to $\psi(N) + d\ \deg(N)$.

In the case where $M = (A/\mathfrak{p})[d]$, $N = A[d']$, and $\mathfrak{p}$ is a height one homogeneous prime, it follows from the fact that $A_\mathfrak{p}$ is a Dedekind domain that $\mathrm{Ext}^1_A(M,N)_\mathfrak{p}$ and $M_\mathfrak{p}$ have the same length over $A_\mathfrak{p}$. So by Lemma 2.4.1 we have $\psi(\mathrm{Ext}^1_A(M,N)) = \mathrm{rank}_A(N)\psi(M)$. We also have $\mathrm{Hom}_A(M,N) = 0$, and so the equation is proved.

In the case where $M = (A/\mathfrak{p})[d]$ and $N = (A/\mathfrak{p}')[d']$ with $\mathfrak{p}$ and $\mathfrak{p}'$ height one homogeneous primes, again all terms in the equation vanish unless $\mathfrak{p} = \mathfrak{p}'$. In this case, $\mathrm{Hom}_A(M,N)_\mathfrak{p}$ and $\mathrm{Ext}^1_A(M,N)_\mathfrak{p}$ have the same length over $A_\mathfrak{p}$, and so by Lemma 2.4.1 we have $\psi(\mathrm{Hom}_A(M,N)) = \psi(\mathrm{Ext}^1_A(M,N))$. So both sides of the equation are zero. □

Corollary 3.3.3 *If A is as in the theorem and M is a finitely generated torsion-free graded A-module, then*

$$\psi(M) + \psi(M^*) = 2\,\mathrm{rank}_A(M)\psi(A).$$

Proof If M is torsion-free then $\mathrm{Ext}^1_A(M,A)$ has Krull dimension at most $n-2$ and so $\psi(\mathrm{Ext}^1_A(M,A)) = 0$. □

3.4 Reflexive modules

The class of modules generalizing the divisorial ideals is the class of reflexive modules.

There is a natural map

$$\phi : M \to M^{**}$$

given by $\phi(m)(f) = f(m)$, and we say that M is **reflexive** if ϕ is an isomorphism. It is easy to see that every map from M to a reflexive module maps through ϕ.

If A is an integral domain, the **torsion** submodule M_0 of M consists of the elements $m \in M$ for which there exists a non-zero element $a \in A$ with $am = 0$. The module M is **torsion-free** if the torsion submodule is zero. Note that M/M_0 is always torsion-free. If $f : M \to A$ is any A-module homomorphism, then f takes the value zero on the torsion submodule. In particular, M^* is always torsion-free, and so every reflexive A-module is torsion-free.

Lemma 3.4.1 *Suppose that A is a normal domain, with field of fractions L, and M is a finitely generated A-module. Then*

(i) *$M^* = \bigcap_{\mathfrak{p}} M^*_{\mathfrak{p}}$, where $\mathfrak{p}$ runs over the prime ideals of height one in A, and the intersection is taken in $L \otimes_A M^*$.*

(ii) *If M is torsion-free then ϕ induces an isomorphism $M^{**} \cong \bigcap_{\mathfrak{p}} M_{\mathfrak{p}}$ ($\mathfrak{p}$ as before). In particular, M is reflexive if and only if $M = \bigcap_{\mathfrak{p}} M_{\mathfrak{p}}$. A fractional ideal $\mathfrak{a}$ of A is divisorial if and only if it is reflexive as an A-module.*

(iii) *The kernel of $\phi : M \to M^{**}$ is equal to the torsion submodule M_0 of M.*

(iv) *M^* is reflexive.*

(v) *If M_1 and M_2 are finitely generated A-modules, then*

$$\mathrm{Hom}_A(M_1, M_2)^{**} \cong \mathrm{Hom}_A(M_1^{**}, M_2^{**}).$$

Proof (i) If $f \in \bigcap_{\mathfrak{p}} M^*_{\mathfrak{p}}$ then for all $m \in M$, $f(m) \in \bigcap_{\mathfrak{p}} A_{\mathfrak{p}} = A$ by Corollary 3.2.6, and so $f \in M^*$.

(ii) If $\mathfrak{p}$ is a height one prime ideal in A, then $A_{\mathfrak{p}}$ is a principal ideal domain and so torsion-free modules are free and hence reflexive. So ϕ induces an isomorphism

$$M^{**} = \bigcap_{\mathfrak{p}} M^{**}_{\mathfrak{p}} \cong \bigcap_{\mathfrak{p}} M_{\mathfrak{p}}.$$

The statement about fractional ideals follows from Corollary 3.2.6.

(iii) M_0 is in the kernel of ϕ, and by (ii), $M/M_0 \to M^{**}$ is injective.

(iv) By (i) and (iii), ϕ induces an isomorphism

$$M^{***} \cong \bigcap_{\mathfrak{p}} M^*_{\mathfrak{p}} = M^*.$$

(v) Replacing M_1 and M_2 by the quotients by their torsion submodules does not affect either side of this isomorphism, so without loss of generality M_1 and M_2 are torsion-free. So

$$\begin{aligned}\mathrm{Hom}_A(M_1, M_2)^{**} &\cong \bigcap_{\mathfrak{p}} \mathrm{Hom}_A(M_1, M_2)_{\mathfrak{p}} = \mathrm{Hom}_A(M_1, \bigcap_{\mathfrak{p}} (M_2)_{\mathfrak{p}}) \\ &\cong \mathrm{Hom}_A(M_1, M_2^{**}) \cong \mathrm{Hom}_A(M_1^{**}, M_2^{**}),\end{aligned}$$

where the intersections are taken in $\mathrm{Hom}_L(L \otimes_A M_1, L \otimes_A M_2)$. □

3.5 Divisor classes and unique factorization

Among the divisors, those represented by principal fractional ideals are called **principal divisors**. These form a subgroup of $D(A)$ denoted $F(A)$. If we denote by $U(A)$ the group of units in A, then the map sending a non-zero element of the field of fractions L to the principal fractional ideal it generates gives an isomorphism $F(A) \cong L^\times / U(A)$.

The group $D(A)/F(A)$ is written $\mathrm{Cl}(A)$ and called the **divisor class group** of A. Thus we have an exact sequence

$$0 \to U(A) \to L^\times \to D(A) \to \mathrm{Cl}(A) \to 0.$$

Proposition 3.5.1 *Suppose that A is a normal domain. Then A is a unique factorization domain if and only if $\mathrm{Cl}(A) = \{0\}$; in other words if and only if every prime ideal of height one is principal.*

Proof If A is a unique factorization domain, then the intersection of an arbitrary collection of principal ideals is principal. So every divisorial fractional ideal is principal and $\mathrm{Cl}(A) = \{0\}$.

Conversely, if $\mathrm{Cl}(A) = \{0\}$ then $D(A) \cong L^\times/U(A)$. So $L^\times/U(A)$ is an ordered abelian group in which every pair of elements has a least upper bound. So we may apply Lemma 3.1.3 to see that the irreducibles in A are prime, so that A is a unique factorization domain. □

3.6 The Picard group

We should stress at this stage that divisorial fractional ideals are not at all the same as invertible ones. We say that a fractional ideal $\mathfrak{a}$ of a normal domain A is **invertible** if $\mathfrak{a}\mathfrak{a}^{-1} = A$. Clearly every principal fractional ideal is invertible, and every invertible fractional ideal is divisorial.

We define the **Picard group** $\mathrm{Pic}(A)$ to be the quotient of the invertible fractional ideals in $D(A)$ by the principal fractional ideals $F(A)$, so that $\mathrm{Pic}(A) \subseteq \mathrm{Cl}(A)$. The following theorem should be contrasted with Theorem 3.9.2.

Theorem 3.6.1 (Kang [53]) $\mathrm{Pic}(K[V]^G) = 0$.

Proof It suffices to show that if $\mathfrak{a}$ is an ideal in $K[V]^G$ with $\mathfrak{a}\mathfrak{a}^{-1} = K[V]^G$ then $\mathfrak{a}$ is principal. Since $K[V]$ is a principal ideal domain, $\mathfrak{a}K[V]$ is principal, say $\mathfrak{a}K[V] = xK[V]$. Then for any $g \in G$, the ideals $g(x)K[V]$ and $xK[V]$ are equal, so that $g(x)/x$ is a unit in $K[V]$, and is hence a scalar. So for some non-zero $\lambda_g \in K$ we have $g(x) = \lambda_g x$.

Choose elements $a_1, \dots, a_s \in \mathfrak{a}$ and $a'_1, \dots, a'_s \in \mathfrak{a}^{-1}$ with $\sum_{i=1}^s a_i a'_i = 1$ in $K[V]^G$. Set $\alpha_i = a_i x^{-1}$ and $\alpha'_i = a'_i x$, so that $\sum_{i=1}^s \alpha_i \alpha'_i = 1$ and the α_i and α'_i are in $K[V]$. Not all the α_i can be in the ideal $K[V]_+$ generated by the elements of positive degree, say $\alpha_1 \notin K[V]_+$. Then for each $g \in G$ the equation $g(\alpha_1) = \lambda_g^{-1}\alpha_1$ forces $\lambda_g = 1$, and so $x \in K[V]^G$ and $\mathfrak{a} = xK[V]^G$. □

3.7 The trace

Suppose that L' is a finite extension field of L. Regarding L' as a vector space over L, we can choose a basis and write multiplication by an element $x \in L'$ as a matrix with entries in L. The trace of this matrix is written $\mathrm{Tr}_{L'/L}(x)$, and is independent of the choice of basis. If $y \in L$, then multiplication by y is given by a diagonal matrix, and so

$$\mathrm{Tr}_{L'/L}(xy) = \mathrm{Tr}_{L'/L}(x)y.$$

If $L'' \supseteq L' \supseteq L$ then the trace is transitive in the sense that

$$\mathrm{Tr}_{L'/L} \circ \mathrm{Tr}_{L''/L'} = \mathrm{Tr}_{L''/L}.$$

Lemma 3.7.1 *If $L' \supseteq L$ is Galois (i.e., normal and separable) with Galois group G, then $\mathrm{Tr}_{L'/L}(x) = \sum_{g \in G} g(x)$.*

Proof By the normal basis theorem, there is an element $y \in L'$ such that as g_i runs over the elements of G, $g_i(y)$ forms a basis of L' as an L-vector space. Suppose that $xg_i(y) = \sum_j a_{ij} g_j(y)$. Then a_{jj} is equal to the coefficient of $g_j(y)$ in $xg_j(y)$ with respect to this basis, which is the same as the coefficient of y in $g_j^{-1}(x)y$. So $\mathrm{Tr}_{L'/L}(x) = \sum_j a_{jj}$ is the coefficient of y in $(\sum_{g \in G} g(x))y$. Since $\sum_{g \in G} g(x)$ is already in L, this is the coefficient. □

Lemma 3.7.2 *$\mathrm{Tr}_{L'/L}$ is identically zero unless L' is a separable extension of L, in which case the bilinear form $(x, y) \mapsto \mathrm{Tr}_{L'/L}(xy)$ is a non-degenerate pairing $L' \otimes_L L' \to L$.*

Proof If $L' \supseteq L$ is purely inseparable with no intermediate fields, then $L' = L(y)$ for some $y \in L'$ with $y^p \in L$ and $p = \mathrm{char}(L)$. Using $1, y, \dots, y^{p-1}$ as a basis for L' over L, we see that the diagonal entries for the action of $x \in L'$ are all equal, so the trace is zero. By transitivity, it follows that the trace is zero unless the extension is separable. On the other hand, if the extension is separable then it may be embedded in a Galois extension, and again by transitivity the trace is non-zero by the previous lemma. Finally, if the trace is non-zero, then given a non-zero element $x \in L'$ there exists $y \in L'$ with $\mathrm{Tr}_{L'/L}(xy) \neq 0$, and so the bilinear form is non-degenerate. □

Proposition 3.7.3 *Suppose that A is a normal domain with field of fractions L. If L' is a finite separable extension of L and B is the integral closure of A in L', then B is finitely generated as an A-module, and hence also a normal domain.*

Proof We first claim that we may choose an L-basis of L' consisting of elements of B. Namely, since B is integral over A, every element $b \in B$ satisfies some equation of the form

$$b^r + a_{r-1}b^{r-1} + \cdots + a_0 = 0$$

with $a_i \in A$. Since B is an integral domain, we may suppose $a_0 \neq 0$. Thus setting $\tilde{b} = -b^{r-1} - a_{r-1}b^{r-2} - \cdots - a_1$, we see that $\tilde{b}b = a_0$ is a non-zero element of A. Let $m = |L' : L|$, and choose a basis $y_1, \dots, y_m$ of L' over L. If $y_i = b_i/b'_i$ with b_i, $b'_i \in B$, then choosing $\tilde{b}_i$ as above, we see that the elements $x_i = \tilde{b}_i b_i y_i = \tilde{b}_i b'_i \in B$ also form a basis of L' over L, so that B contains a free A-module F of rank m generated by the x_i.

Let

$$\mathfrak{D}^{-1}_{B/A} = \{x \in L' \mid \forall y \in B, \mathrm{Tr}_{L'/L}(xy) \in A\}$$

(this is the inverse different, of which we will have more to say in Section 3.10). Then by Lemma 3.7.2, as a B-module we have $\mathfrak{D}^{-1}_{B/A} \cong \mathrm{Hom}_A(B, A)$. Similarly, define

$$F' = \{x \in L' \mid \forall y \in F, \mathrm{Tr}_{L'/L}(xy) \in A\},$$

so that F' is a free A-module of rank m. Then $F' \supseteq \mathfrak{D}^{-1}_{B/A} \supseteq B$, and so B is a submodule of the finitely generated A-module F'. □

3.8 Ramification

Suppose that $A \subseteq B$ is a finite extension of normal domains, with fields of fractions $L \subseteq L'$. If $\mathfrak{p}$ is a prime ideal in B, then $\mathfrak{p} = \mathfrak{P} \cap A$ is a prime ideal in A, and by Theorem 1.4.4, $\mathfrak{P}$ has height one if and only if $\mathfrak{p}$ has height one. If $\mathfrak{p}$ and $\mathfrak{P}$ have height one, then the discussion at the end of Section 3.2 shows that the restriction of $v_{\mathfrak{P}}$ to L is some positive integer multiple of $v_{\mathfrak{p}}$, say

$$v_{\mathfrak{P}} = e(\mathfrak{P}, \mathfrak{p})v_{\mathfrak{p}}.$$

The positive integer $e = e(\mathfrak{P}, \mathfrak{p})$ is called the **ramification index** of $\mathfrak{P}$ over $\mathfrak{p}$. It is characterized by the equation $\mathfrak{p}B_{\mathfrak{P}} = \mathfrak{P}^e B_{\mathfrak{P}}$. We say that the extension $B \supseteq A$ is **unramified** at $\mathfrak{P}$ if $e(\mathfrak{P}, \mathfrak{p}) = 1$ and $B/\mathfrak{P}$ is a separable extension of $A/\mathfrak{p}$.

Let $f = f(\mathfrak{P}, \mathfrak{p})$ be the degree of the field extension $|B_{\mathfrak{P}}/\mathfrak{P}B_{\mathfrak{P}} : A_{\mathfrak{p}}/\mathfrak{p}A_{\mathfrak{p}}|$. If π is an element of $\mathfrak{P}B_{\mathfrak{P}}$ not in $\mathfrak{P}^2 B_{\mathfrak{P}}$, then multiplication by π is an isomorphism $\mathfrak{P}^i B_{\mathfrak{P}} \to \mathfrak{P}^{i+1} B_{\mathfrak{P}}$ for each value of i. Examining the series

$$B_{\mathfrak{P}} \supseteq \mathfrak{P}B_{\mathfrak{P}} \supseteq \cdots \supseteq \mathfrak{P}^e B_{\mathfrak{P}} = \mathfrak{p}B_{\mathfrak{P}},$$

whose quotients are all isomorphic as $A_{\mathfrak{p}}/\mathfrak{p}A_{\mathfrak{p}}$-modules, we see that $B_{\mathfrak{P}}/\mathfrak{p}B_{\mathfrak{P}}$ is a free $A_{\mathfrak{p}}/\mathfrak{p}_A\mathfrak{p}$-module of rank ef. Since $A_{\mathfrak{p}}$ is a local Dedekind domain, it follows that $B_{\mathfrak{P}}$ is a free $A_{\mathfrak{p}}$-module of rank ef.

If $0 \neq x \in L$, then considered as an element of L' we have

$$d(x) = \sum v_{\mathfrak{P}}(x)d(\mathfrak{P}) = \sum v_{\mathfrak{p}}(x)e(\mathfrak{P}, \mathfrak{p})d(\mathfrak{P}).$$

So if we define a map $j : D(A) \to D(B)$ by $j(d(\mathfrak{p})) = \sum e(\mathfrak{P},\mathfrak{p})d(\mathfrak{P})$, then j induces the obvious map from $F(A)$ to $F(B)$, and hence a well defined map $\bar{j} : \mathrm{Cl}(A) \to \mathrm{Cl}(B)$.

Now suppose that $L \subseteq L'$ is a Galois extension with Galois group G. Then by Theorem 1.4.4, $A = B^G$ and for any prime ideal $\mathfrak{p}$ of height one in A, the primes $\mathfrak{P}$ of B lying over $\mathfrak{p}$ are G-conjugate, so $e(\mathfrak{P},\mathfrak{p})$ is independent of $\mathfrak{P}$, and is written $e(\mathfrak{p})$. If $\mathfrak{P} = \mathfrak{P}_1, \dots, \mathfrak{P}_r$ are the primes lying over $\mathfrak{p}$, we write G_d for the setwise stabilizer of $\mathfrak{P}$ in G; this is the **decomposition group**. We have $r = |G : G_d|$. If $\mathfrak{P}_i = g(\mathfrak{P})$ for $g \in G$, then the setwise stabilizer of $\mathfrak{P}_i$ is gG_dg^{-1}. The local Dedekind domain $B_\mathfrak{P}$ is an extension of $A_\mathfrak{p}$ of degree $|G_d|$, and so it follows that $|G_d| = ef$.

We write G_t for the normal subgroup of G_d consisting of those elements which act trivially on $B_\mathfrak{P}/\mathfrak{P}B_\mathfrak{P}$; this is the **inertia group**. If we wish to make the prime explicit in the notation, we write $G_\mathfrak{P}$ instead of G_t. The quotient G_d/G_t acts as Galois automorphisms on $B_\mathfrak{P}/\mathfrak{P}B_\mathfrak{P}$ fixing $A_\mathfrak{p}/\mathfrak{p}A_\mathfrak{p}$, and so $|G_d/G_t|$ divides f, which implies that e divides $|G_t|$. Equality need not hold, since $B_\mathfrak{P}/\mathfrak{P}B_\mathfrak{P}$ need not be a separable field extension of $A_\mathfrak{p}/\mathfrak{p}A_\mathfrak{p}$. However, since ramification indexes are obviously multiplicative over field extensions, to calculate the ramification index it suffices to calculate it over the fixed ring of G_t, which is often simpler than calculating it over the fixed ring of G.

Theorem 3.8.1 *With the above hypotheses, there is an exact sequence*

$$0 \to \ker \bar{j} \to H^1(G, U(B)) \to \bigoplus_{\mathfrak{p}} \mathbb{Z}/e(\mathfrak{p}) \to \operatorname{Coker} \bar{j} \to 0,$$

where $\bar{j}$ is regarded as a map from $\mathrm{Cl}(A)$ to $\mathrm{Cl}(B)^G$, and $\mathfrak{p}$ runs over the height one prime ideals in A.

Proof By Hilbert's Theorem 90, $H^1(G,(L')^\times) = 0$, so taking fixed points of G on the short exact sequence

$$0 \to U(B) \to (L')^\times \to F(B) \to 0$$

we obtain an exact sequence

$$0 \to U(A) \to L^\times \to F(B)^G \to H^1(G, U(B)) \to 0.$$

We also have a short exact sequence

$$0 \to U(A) \to L^\times \to F(A) \to 0.$$

The maps $U(A) \to L^\times$ in these sequences are the same, so we obtain a short exact sequence

$$0 \to F(A) \to F(B)^G \to H^1(G, U(B)) \to 0.$$

The left hand map in this sequence is just the restriction of j to $F(A)$. On the other hand, the description of $D(A)$ and $D(B)$ given in Theorem 3.2.4 makes it clear that we have a short exact sequence

$$0 \to D(A) \xrightarrow{j} D(B)^G \to \bigoplus_{\mathfrak{p}} \mathbf{Z}/e(\mathfrak{p}) \to 0.$$

Applying the snake lemma to the top two rows of the diagram

$$\begin{array}{ccccccc} & 0 & & 0 & & & \\ & \downarrow & & \downarrow & & & \\ 0 \to & F(A) & \to & F(B)^G & \to & H^1(G, U(B)) & \to 0 \\ & \downarrow & & \downarrow & & \downarrow & \\ 0 \to & D(A) & \xrightarrow{j} & D(B)^G & \to & \bigoplus_{\mathfrak{p}} \mathbf{Z}/e(\mathfrak{p}) & \to 0 \\ & \downarrow & & \downarrow & & & \\ & \mathrm{Cl}(A) & \xrightarrow{\bar{j}} & \mathrm{Cl}(B)^G & & & \\ & \downarrow & & \downarrow & & & \\ & 0 & & 0 & & & \end{array}$$

we see that the kernel and cokernel of $H^1(G, U(B)) \to \bigoplus_{\mathfrak{p}} \mathbf{Z}/e(\mathfrak{p})$ are equal to the kernel and cokernel of $\bar{j} : \mathrm{Cl}(A) \to \mathrm{Cl}(B)^G$. □

3.9 $\mathrm{Cl}(K[V]^G)$

We apply the theory described in the last section in the situation where $A = K[V]^G$, $B = K[V]$. Since $K[V]$ is a unique factorization domain, we have $\mathrm{Cl}(K[V]) = 0$, so that $\ker \bar{j} = \mathrm{Cl}(K[V]^G)$ and $\mathrm{Coker}\, \bar{j} = 0$. Moreover, $U(B) = K^\times$ with trivial G-action, and so the exact sequence of Theorem 3.8.1 reduces to

$$0 \to \mathrm{Cl}(K[V]^G) \to \mathrm{Hom}(G, K^\times) \to \bigoplus_{\mathfrak{p}} \mathbf{Z}/e(\mathfrak{p}) \to 0.$$

If W is a codimension one subspace of V, we write $\mathfrak{P}_W$ for the corresponding height one prime ideal in $K[V]$. It is generated by any non-zero linear form vanishing on W, and in particular it is homogeneous. We write G_W for the pointwise stabilizer in G of W, and recall (Section 2.6) that W is said to be a reflecting hyperplane if $G_W \neq 1$.

Lemma 3.9.1 *Suppose that G acts faithfully on V. If $\mathfrak{P}$ is a prime ideal of height one in $K[V]$ with non-trivial inertia group, then $\mathfrak{P} = \mathfrak{P}_W$ for some reflecting hyperplane $W \subset V$. The ramification group is equal to G_W. Setting $\mathfrak{p} = \mathfrak{P} \cap K[V]^G$, the ramification index $e(\mathfrak{p})$ is equal to $|G_W|$ if $\operatorname{char}(K) = 0$, and the p'-part $|G_W|_{p'}$ if $\operatorname{char}(K) = p$.*

Proof Since $K[V]$ is a unique factorization domain, the ideal $\mathfrak{P}$ is generated by a homogeneous element of some degree d. If $d \geq 2$ then any element of the inertia group fixes the degree one elements of $K[V]$, and since the action is faithful it is thus the identity element. So $\mathfrak{P}$ is generated by a degree one element. The inertia group of $\mathfrak{p}$ is equal to the pointwise stabilizer of the corresponding hyperplane W.

Now if $\operatorname{char}(K) = 0$ then G_W is cyclic, generated by an element g of order h acting diagonally with a single eigenvalue other than one. So there is a basis $v_1, \dots, v_n$ of V with $g(v_i) = v_i$ $(1 \leq i \leq n)$ and $g(v_n) = \lambda v_n$ $(\lambda^h = 1)$. If $x_1, \dots, x_n$ is the dual basis of V^*, then $K[V]^{G_W} = K[x_1, \dots, x_{n-1}, x_n^h]$. We have $\mathfrak{p} = (x_n^h)$, and $e(\mathfrak{p}) = h$.

On the other hand, if $\operatorname{char}(K) = p$ then $|G_W| = p^a.h$, and G_W is a split extension. It has a normal elementary abelian p-subgroup $E_W = O_p(G_W)$ generated by elements $g_1, \dots, g_a$, and a cyclic complement generated by an element g of order h coprime to p acting diagonally with a single eigenvalue other than one. There is a basis $v_1, \dots, v_n$ of V with $g(v_i) = v_i$ $(1 \leq i \leq n-1)$, $g(v_n) = \lambda v_n$ $(\lambda^h = 1)$, and $g_j(v_i) = v_i$ $(1 \leq i \leq n-1)$, $g_j(v_n) = v_n + \sum_{i=1}^{n-1} \lambda_{ij} v_i$ for suitable constants $\lambda_{ij} \in K$. If $x_1, \dots, x_n$ is the dual basis of V^*, then $g(x_i) = x_i$ $(1 \leq i \leq n)$, $g(x_n) = \lambda^{-1} x_n$, $g_j(x_i) = x_i - \lambda_{ij} x_n$ $(1 \leq i \leq n-1)$ and $g_j(x_n) = x_n$. Again we have $\mathfrak{p} = (x_n^h)$ and $e(\mathfrak{p}) = h$. □

The homomorphism

$$\operatorname{Hom}(G, K^\times) \to \bigoplus_W \mathbb{Z}/|G_W|_{p'}$$

may be described as follows. If $\phi : G \to K^\times$ is a group homomorphism and W is a reflecting hyperplane, then ϕ has $E_W = O_p(G_W)$ in its kernel, and so induces a homomorphism from G_W/E_W to $K^\times$. So we get an element of

$$\operatorname{Hom}(G_W, K^\times) = \operatorname{Hom}(G_W/E_W, K^\times) \cong \mathbb{Z}/|G_W|_{p'}.$$

This proves the following formula for $\operatorname{Cl}(K[V]^G)$.

Theorem 3.9.2 *$\operatorname{Cl}(K[V]^G)$ is isomorphic to the subgroup of $\operatorname{Hom}(G, K^\times)$ consisting of those homomorphisms which take the value one on every pseudoreflection.* □

Corollary 3.9.3 (Nakajima) *$K[V]^G$ is a unique factorization domain if and only if there are no non-trivial homomorphisms $G \to K^\times$ taking the value one on every pseudoreflection.* □

3.10 The different

Suppose that $A \subseteq B$ is a finite extension of normal domains, with fields of fractions $L \subseteq L'$ a separable extension. There are two distinct notions in this situation which generalize the usual different from algebraic number theory. Following Auslander and Buchsbaum [10], we shall call these the different and the homological different. In the case where B is a projective A-module, the two notions coincide. This fact will be proved in the next section, in preparation for the discussion of groups generated by pseudoreflections in Section 7.2.

We begin with the different. Since $L \subseteq L'$ is separable, by Lemma 3.7.2 the trace form $(x, y) \mapsto \mathrm{Tr}_{L'/L}(xy)$ is a non-degenerate pairing $L' \otimes_L L' \to L$. We set

$$\mathfrak{D}_{B/A}^{-1} = \{x \in L' \mid \forall y \in B,\ \mathrm{Tr}_{L'/L}(xy) \in A\}.$$

It is convenient to write this condition as $\mathrm{Tr}_{L'/L}(xB) \subseteq A$.

The map $\mathfrak{D}_{B/A}^{-1} \to B^* = \mathrm{Hom}_A(B, A)$ which sends x to the map $y \mapsto \mathrm{Tr}_{L'/L}(xy)$ is an isomorphism. So by Lemma 3.4.1, $\mathfrak{D}_{B/A}^{-1}$ is a reflexive A-module. It follows from that lemma together with Theorem 1.4.4 that a finitely generated B-module which is reflexive as an A-module is also reflexive as a B-module, so $\mathfrak{D}_{B/A}^{-1}$ is a divisorial fractional ideal, called the **inverse different**. If A and B are graded in such a way that the inclusion $A \hookrightarrow B$ preserves the grading, then L' has a basis over L consisting of homogeneous elements. So $\mathrm{Tr}_{L'/L} : B \to A$ preserves the grading, and Lemma 3.3.1 shows that $\mathfrak{D}_{B/A}^{-1}$ comes with a grading in such a way that the inclusion $B \hookrightarrow \mathfrak{D}_{B/A}^{-1}$ is a graded map.

The inverse fractional ideal to $\mathfrak{D}_{B/A}^{-1}$,

$$\mathfrak{D}_{B/A} = (\mathfrak{D}_{B/A}^{-1})^{-1}$$

is called the **different**. The map $\mathrm{Hom}_B(B^*, B) \to \mathfrak{D}_{B/A}$ which sends ρ to $\rho(\mathrm{Tr}_{L'/L})$ is an isomorphism of B-modules. Since $\mathfrak{D}_{B/A}^{-1} \supseteq B$, we have $\mathfrak{D}_{B/A} \subseteq B$ and so the different is a divisorial ideal in B. If A and B are graded in such a way that the inclusion $A \hookrightarrow B$ preserves the grading, then another application of Lemma 3.3.1 shows that $\mathfrak{D}_{B/A}$ is a homogeneous divisorial ideal in B.

Lemma 3.10.1 *Suppose that $A \subseteq B \subseteq C$ are finite extensions of normal domains, with fields of fractions $L \subseteq L' \subseteq L''$ separable. Then*

$$\mathfrak{D}_{C/A} = \overline{\mathfrak{D}_{C/B}\mathfrak{D}_{B/A}}.$$

Proof By transitivity of the trace, an element x of L'' lies in $\mathfrak{D}_{C/A}^{-1}$ if and only if

$$\mathrm{Tr}_{L'/L}\mathrm{Tr}_{L''/L'}(xC) = \mathrm{Tr}_{L''/L}(xC) \subseteq A.$$

This happens if and only if $\mathrm{Tr}_{L'/L}(B\mathrm{Tr}_{L''/L'}(xC)) \subseteq A$, since we may take elements of B inside the second trace. This is the same as saying that $\mathrm{Tr}_{L''/L'}(xC) \subseteq \mathfrak{D}_{B/A}^{-1}$, or equivalently $\mathrm{Tr}_{L''/L'}(xC\mathfrak{D}_{B/A}) \subseteq B$, or $x\mathfrak{D}_{B/A} \subseteq \mathfrak{D}_{C/B}^{-1}$. This proves that

$$\mathfrak{D}_{C/A}^{-1} = (\mathfrak{D}_{C/B}\mathfrak{D}_{B/A})^{-1},$$

from which the lemma follows. □

Theorem 3.10.2 *Suppose that $A \subseteq B$ is a finite extension of normal domains, with fields of fractions $L \subseteq L'$ a separable extension. If $\mathfrak{P}$ and $\mathfrak{p} = A \cap \mathfrak{P}$ are prime ideals of height one in B and A, then $e(\mathfrak{P}, \mathfrak{p}) > 1$ if and only if $\mathfrak{D}_{B/A} \subseteq \mathfrak{P}$.*

Proof It follows easily from the definition of $\mathfrak{D}_{B/A}$ that if we form the localization $S^{-1}\mathfrak{D}_{B/A}$ by inverting the set S of all elements of A not in $\mathfrak{p}$, the result is equal to $\mathfrak{D}_{S^{-1}B/S^{-1}A}$. So we can replace A and B by $S^{-1}A$ and $S^{-1}B$, so that they have Krull dimension one, and are hence Dedekind domains. The statement is then a standard theorem in algebraic number theory, see for example Cassels and Fröhlich [25], p. 20–22, or Serre [92], §III.5. □

Proposition 3.10.3 *Suppose that $L \subseteq L'$ is a Galois extension. Then we can calculate the different locally as follows. If $\mathfrak{P}$ is a height one prime in B with inertia group $G_{\mathfrak{P}}$, then*

$$v_{\mathfrak{P}}(\mathfrak{D}_{B/A}) = v_{\mathfrak{P}}(\mathfrak{D}_{B/B^{G_{\mathfrak{P}}}}).$$

Proof Set $\mathfrak{Q} = \mathfrak{P} \cap B^{G_{\mathfrak{P}}}$. Then $B^{G_{\mathfrak{P}}}/A$ is unramified at $\mathfrak{Q}$, which implies that $\mathfrak{D}_{B^{G_{\mathfrak{P}}}/A} \not\subseteq \mathfrak{Q}$, so that $B\mathfrak{D}_{B^{G_{\mathfrak{P}}}/A}^{-1} \not\subseteq \mathfrak{P}$. The proposition now follows from Lemma 3.10.1. □

3.11 The homological different

We now describe the homological different. Denote by $\mathfrak{J}$ the kernel of the multiplication map $\phi : B \otimes_A B \to B$, namely the ideal generated by the elements $x \otimes 1 - 1 \otimes x$. Denote by $\mathfrak{N}$ the annihilator in $B \otimes_A B$ of $\mathfrak{J}$. Then the **homological different** is defined to be $\mathfrak{H}_{B/A} = \phi(\mathfrak{N})$.

If A and B are finitely generated commutative algebras over an algebraically closed field K, this has the following geometric description. The inclusion $A \to B$ corresponds to a finite (surjective) map of irreducible varieties $p : V \to W$, where V and W are the maximal ideal spectra of B and A respectively. Then the maximal ideal spectrum of $B \otimes_A B$ is the variety

$$\Delta_W(V) = \{(v_1, v_2) \in V \times V \mid p(v_1) = p(v_2)\}.$$

The map ϕ corresponds to the inclusion of the diagonal

$$\Delta(V) = \{(v, v) \in V \times V\}$$

which is an irreducible component of $\Delta_W(V)$. Thus $\mathfrak{J}$ is the ideal of functions on $\Delta_W(V)$ vanishing on the remaining components, and $\mathfrak{H}_{B/A}$ is the ideal in B consisting of the functions which vanish on the intersection of these remaining components with $\Delta(V)$. This subvariety is called the **branch locus.**

If, furthermore, A is the fixed points of a faithful action of a finite group G on B, then W is the quotient variety V/G (see the remarks at the end of Section 1.4). In this case,

$$\Delta_W(V) = \{(v_1, v_2) \in V \times V \mid \exists g \in G,\ g(v_1) = v_2\}$$

is the union of $|G|$ diagonals $\{(v, g(v)) \in V \times V\}$, and the locus of ramification is precisely the set of points in V with a non-trivial stabilizer.

Theorem 3.11.1 (Auslander–Buchsbaum) *We have $\mathfrak{H}_{B/A} \subseteq \mathfrak{D}_{B/A}$, and if B is projective as an A-module then $\mathfrak{H}_{B/A} = \mathfrak{D}_{B/A}$.*

Proof Set $(L')^* = \mathrm{Hom}_L(L', L)$ and $B^* = \mathrm{Hom}_A(B, A)$. Define a map

$$\sigma' : L' \otimes_L L' \to \mathrm{Hom}_L((L')^*, L')$$

by $\sigma'(x \otimes y)(f) = xf(y)$. Since L' is a finite dimensional vector space over L, σ' is an isomorphism. Its inverse is given by $(\sigma')^{-1}(\rho)(f) = \sum_j \rho(x_j^*) \otimes x_j$, where x_j and x_j^* are dual bases of L' and $(L')^*$ over L. Furthermore, σ' restricts to give a map

$$\sigma : B \otimes_A B \to \mathrm{Hom}_A(B^*, B),$$

which is an isomorphism provided B is projective as an A-module.

Now consider the diagram

$$\begin{array}{ccc}
\mathrm{Hom}_B(B^*, B) & \to & \mathrm{Hom}_{L'}((L')^*, L') \\
\downarrow i & & \downarrow i' \\
\mathrm{Hom}_A(B^*, B) & \to & \mathrm{Hom}_L((L')^*, L') \\
\uparrow \sigma & & \cong \uparrow \sigma' \\
B \otimes_A B & \to & L' \otimes_L L' \\
\downarrow \phi & & \downarrow \phi' \\
B & \to & L'
\end{array}$$

where i and i' are the inclusions and ϕ, ϕ' are given by multiplication. In the top line, L' acts on $(L')^*$ by multiplying before taking the effect of the map. Now $\mathrm{Tr}_{L'/L}(x) = \sum_j x_j^*(xx_j)$, so if $\rho \in \mathrm{Hom}_{L'}((L')^*, L')$ we have $\rho(\mathrm{Tr}_{L'/L}) = \sum_j x_j\rho(x_j^*)$. So the composite down the right hand side of the diagram is given by

$$\phi'(\sigma')^{-1}i'(\rho) = \phi'(\sum_j \rho(x_j^*) \otimes x_j) = \sum x_j\rho(x_j^*) = \rho(\mathrm{Tr}_{L'/L}).$$

It follows that the image of $\mathrm{Hom}_B(B^*, B)$ under $\phi'(\sigma')^{-1}i'$ is $\mathfrak{D}_{B/A}$.

We make $\mathrm{Hom}_A(B^*, B)$ into a $B \otimes_A B$-module via $((x \otimes y)\rho)(f) = x\rho(f \circ y)$, where the y on the right hand side of this denotes the map from B to itself given by multiplication by y. Thus σ is a $B \otimes_A B$-module homomorphism. An element $\rho \in \mathrm{Hom}_A(B^*, B)$ is in the image of i if and only if it is a B-module homomorphism; namely if and only if $x \otimes 1$ and $1 \otimes x$ act on ρ in the same way. This is the same as saying that $\mathfrak{J}\rho = 0$. Since $\mathfrak{N}$ is the annihilator of $\mathfrak{J}$, this implies that $\sigma(\mathfrak{N})$ lies in the image of i. Passing to the right hand side of the diagram and applying $\phi'(\sigma')^{-1}$, it follows that $\mathfrak{H}_{B/A} = \phi(\mathfrak{N}) \subseteq \mathfrak{D}_{B/A}$.

Finally, if B is projective as an A-module, then σ is an isomorphism and $\sigma(\mathfrak{N})$ is equal to the image of i, so that $\mathfrak{H}_{B/A} = \mathfrak{D}_{B/A}$. □

3.12 A ramification formula

Suppose that $A = \bigoplus_{j=0}^{\infty}$ is a graded normal domain with $A_0 = K$, and with field of fractions L. Suppose that L'/L is a finite separable field extension. Let B be the integral closure of A in L', and suppose that B is graded in such a way that the inclusion $A \hookrightarrow B$ preserves degrees. Since $\mathrm{rank}_A(B) = |L' : L|$ we have $\deg(B) = |L' : L| \deg(A)$. We have the following ramification formula (Benson and Crawley–Boevey [14]) for the invariant ψ introduced in Section 2.4:

Theorem 3.12.1 $|L' : L|\psi(A) - \psi(B) = \frac{1}{2}\sum_{\mathfrak{P}} v_{\mathfrak{P}}(\mathfrak{D}_{B/A})\psi(B/\mathfrak{P})$, *where the sum runs over the homogeneous height one primes.*

Proof By Lemma 2.4.1, we have

$$\begin{aligned}\psi(\mathfrak{D}_{B/A}^{-1}) - \psi(B) = \psi(\mathfrak{D}_{B/A}^{-1}/B) &= \sum_{\mathfrak{P}} \mathrm{length}_{B_{\mathfrak{P}}}((\mathfrak{D}_{B/A}^{-1}/B)_{\mathfrak{P}})\psi(B/\mathfrak{P}) \\ &= \sum_{\mathfrak{P}} v_{\mathfrak{P}}(\mathfrak{D}_{B/A})\psi(B/\mathfrak{P}).\end{aligned}$$

On the other hand, as an A-module we have $\mathfrak{D}_{B/A}^{-1} \cong B^*$, so by Corollary 3.3.3 we have

$$\psi(\mathfrak{D}_{B/A}^{-1}) + \psi(B) = 2|L' : L|\psi(A).$$

Combining these equations proves the theorem. □

We wish to apply the theorem in the following situation. Let $B = \sum_{j=0}^{\infty} B_j$ be a graded normal domain with $B_0 = K$, and with field of fractions L'. If G is a finite group of automorphisms of B preserving the grading, then the theorem applies to the inclusion of $A = B^G$ in B, with the inherited grading.

Corollary 3.12.2 *Suppose that $G_{\mathfrak{P}} \cap G_{\mathfrak{P}'} = 1$ for distinct homogeneous height one prime ideals $\mathfrak{P}$ and $\mathfrak{P}'$. Then*

$$|G|\psi(B^G) - \psi(B) = \sum_{\mathfrak{P}} \left(|G_{\mathfrak{P}}|\psi(B^{G_{\mathfrak{P}}}) - \psi(B)\right)$$

where the sum runs over homogeneous height one primes.

Proof The assumption on the inertia subgroups implies that $\mathfrak{P}$ is the only height one prime in B which ramifies over $B^{G_{\mathfrak{P}}}$, so by applying Theorem 3.12.1 to the action of $G_{\mathfrak{P}}$ on B, we have

$$|G_{\mathfrak{P}}|\psi(B^{G_{\mathfrak{P}}}) - \psi(B) = \tfrac{1}{2}v_{\mathfrak{P}}(\mathfrak{D}_{B/B^{G_{\mathfrak{P}}}})\psi(B/\mathfrak{P}) = \tfrac{1}{2}v_{\mathfrak{P}}(\mathfrak{D}_{B/A})\psi(B/\mathfrak{P}).$$

A second application of the theorem gives the result. □

3.13 The Carlisle–Kropholler conjecture

In this section, we apply Corollary 3.12.2 to the case of polynomial invariants, and prove the conjecture of Carlisle and Kropholler mentioned at the end of the previous chapter.

Suppose that G is a finite group acting faithfully on a vector space V over $\mathbb{F}_p$, and set $B = \mathbb{F}_p[V] = \mathbb{F}_p[x_1, \dots, x_n]$. Then we have $p(B,t) = 1/(1-t)^n$, and so $\psi(B) = 0$. Let $A = B^G$, the ring of invariants.

If W is a reflecting hyperplane, we may choose a basis $v_1, \dots, v_n$ for V in such a way that W is the subspace generated by $v_2, \dots, v_n$. Then the matrices for the action of G_W have the form

$$\begin{pmatrix} * & 0 & \cdots & 0 \\ * & 1 & & 0 \\ \vdots & & & \vdots \\ * & 0 & \cdots & 1 \end{pmatrix}$$

So G_W is a split extension with normal elementary abelian subgroup $(\mathbb{Z}/p)^{a_W}$ ($a_W \le n-1$) and quotient $\mathbb{Z}/h_W$ with h_W a divisor of $p-1$. After a suitable adjustment of the basis, G is generated by elements g and $g_2, \dots, g_{a_W+1}$ acting via

$$\begin{aligned} g(v_1) &= \lambda v_1 & g_i(v_1) &= v_1 + v_i \\ g(v_j) &= v_j & g_i(v_j) &= v_j \qquad (j > 1), \end{aligned}$$

where λ is a primitive h_Wth root of unity in $\mathbb{F}_p$. The dual action on $x_1, \ldots, x_n$ is given by

$$\begin{array}{ll} g(x_1) = \lambda^{-1} x_1 & g_i(x_1) = x_1 \\ g(x_j) = x_j & g_i(x_j) = x_j - \delta_{ij} x_1 \quad (j > 1) \end{array}$$

Here, δ_{ij} denotes the Kronecker delta.

Lemma 3.13.1 *The invariants $B^{G_W} = \mathbb{F}_p[x_1, \ldots, x_n]^{G_W}$ form a polynomial ring*

$$\mathbb{F}_p[x_1^{h_W}, x_2^p - x_2 x_1^{p-1}, \ldots, x_{a_W+1}^p - x_{a_W+1} x_1^{p-1}, x_{a_W+2}, \ldots, x_n].$$

In particular, the Poincaré series of this ring is given by

$$p(B^{G_W}, t) = \frac{1}{(1 - t^{h_W})(1 - t^p)^{a_W}(1 - t)^{n - a_W - 1}}$$

and so

$$\psi(B^{G_W}) = \tfrac{1}{2}((p-1)a_W + h_W - 1)/(h_W p^{a_W}).$$

Proof The given elements are clearly invariant, since they are products of the orbits of the degree one generators. Furthermore, it is clear for the same reason that B is integral over this subring. At the level of fields of fractions, the extension has the right degree, and is hence right by Galois theory. Finally, this polynomial subring is integrally closed, and therefore equal to the invariants. □

The following theorem is the Carlisle–Kropholler conjecture [21], and the proof we have given is due to Benson and Crawley-Boevey [14].

Theorem 3.13.2 $\psi(\mathbb{F}_p[V]^G) = \frac{1}{2|G|} \sum_W ((p-1)a_W + h_W - 1)$

Proof This follows from Corollary 3.12.2 and the above lemma. □

We remark that for larger fields of characteristic p, the invariants of G_W need not form a polynomial ring, so the local calculation of ψ is harder, and in particular depends on more than just a knowledge of $|G_W|$.

Chapter 4

Homological Properties of Invariants

4.1 Minimal resolutions

Usually, the ring of invariants $K[V]^G$ is not a polynomial ring (see Section 7.2). Thus a set of generators for the invariants will usually satisfy some relations. A **syzygy of the first kind** is a polynomial relation between the generating invariants. Similarly, a **syzygy of the second kind** is a polynomial relation between the syzygies of the first kind, and so on. Hilbert's syzygy theorem [45] (see Theorem 4.2.2 and Corollary 4.2.3) states that for each value of r, the syzygies of the rth kind are finitely generated as a module over the polynomial ring on the generating invariants, and that for large enough r, there are no syzygies of the rth kind. In coprime characteristic, we shall see (Corollary 4.4.6) that if $\dim_K V = n$ and the number of generating invariants is s, then there are syzygies of the tth kind if and only if $t \leq s - n$.

The homological interpretation of the Hilbert syzygy theorem is as follows. Write $K[V]^G = K[y_1, \ldots, y_s]/I$, where the y_i are homogeneous generators of degrees k_i, and I is a homogeneous ideal. Then writing A for $K[y_1, \ldots, y_s]$, $K[V]^G$ has a free resolution as an A-module, of the form

$$\cdots \to A\psi_1 \oplus \cdots \oplus A\psi_{t'} \to A\theta_1 \oplus \cdots \oplus A\theta_t \to A \to K[V]^G \to 0.$$

The θ_i are syzygies of the first kind, the ψ_j are syzygies of the second kind, and so on.

In this context, it is no longer necessary to restrict oneself to discussing invariants. The syzygy theorem says more generally that if M is a finitely generated graded module for $A = K[y_1, \ldots, y_s]$ (note that the grading on A puts y_i in degree k_i and not necessarily in degree one) then M has a resolution of finite length (in fact length at most s) by finitely generated free A-modules.

We begin with a discussion of minimal resolutions. For the rest of this section, we retain the notation $A = K[y_1, \dots, y_s]$ as above.

Lemma 4.1.1 *A graded A-module M is projective if and only if it is free.*

Remark In fact, the same remains true of finitely generated ungraded modules over a polynomial ring, as was conjectured by Serre [89] and proved independently by Quillen [83] and Suslin.

Proof of Lemma Suppose that M is projective. Let M_d be the first non-zero graded piece of M, and let $\{m_\alpha\}$ be a basis of M_d as a K-vector space. Then we can write M as a quotient of a free module

$$F \oplus \bigoplus_\alpha A.m_\alpha \to M \to 0$$

where F is a free module generated by elements of degree at least $d+1$. Since M is projective, this map splits. The composite of a splitting with the projection

$$M \to F \oplus \bigoplus_\alpha A.m_\alpha \to \bigoplus_\alpha A.m_\alpha$$

is an isomorphism in degree d, so the generators m_α are in the image, and hence the map is an epimorphism. Since $\bigoplus_\alpha A.m_\alpha$ is projective, this says

$$M = \bigoplus_\alpha A.m_\alpha \oplus M'$$

where the first non-zero graded piece of M' has degree at least $d+1$. Now M' is projective, so we may continue

$$M = \bigoplus_\alpha A.m_\alpha \oplus \bigoplus_\beta A.m'_\beta \oplus \cdots$$

In any particular degree, we only need add a finite number of terms of this expansion, and so M is equal to this infinite direct sum. □

The following proposition guarantees the existence of "projective covers" in this situation.

Proposition 4.1.2 *If M is a graded A-module, let $F_M = A \otimes_K (M/A^+M)$, a free A-module on the graded K-vector space M/A^+M of generators of M. Then there is a surjective map $\rho_M : F_M \twoheadrightarrow M$. If $\rho : F \to M$ is another free module surjecting onto M then $F \cong F' \oplus F_M$ with $\rho|_{F'} = 0$ and $\rho|_{F_M} = \rho_M$.*

Proof Consider the map $F_M \to M/A^+M$ induced by $A \to A/A^+ = K$. Since F_M is free, this lifts to a map $\rho_M : F_M \to M$. Let M' be the image of this map. Then

$$(M/M')/A^+(M/M') = M/(M' + A^+M) = 0,$$

so that $M/M' = 0$ and hence ρ_M is surjective.

Now suppose that $\rho : F \to M$ is another free module surjecting onto M. Since F is free, we can lift to a map $F \to F_M$. Now F surjects onto M/A^+M, and hence onto F_M/A^+F_M and so $F \to F_M$ is surjective. Since F_M is free, we can write $F \cong F' \oplus F_M$ with $\rho|_{F'} = 0$ and $\rho|_{F_M} = \rho_M$. □

Corollary 4.1.3 *If M is a graded A-module then M has a unique (up to isomorphism of chain complexes) minimal projective (= free) resolution*

$$\cdots \to F_1 \to F_0 \to M \to 0.$$

Every projective resolution of M is a direct sum of this one and a split exact sequence of free modules.

Proof This follows by applying the proposition and induction. □

4.2 Hilbert's syzygy theorem

We now wish to show that the minimal resolution over a polynomial ring described in Corollary 4.1.3 has finite length, at most equal to the number s of variables. For this purpose, we introduce the **Koszul complex**.

Given an element x in a commutative ring A, we form a complex $\mathfrak{K}(x) = \mathfrak{K}^A(x)$ as follows. In degrees zero and one, we take a free A-module on a single generator, named 1 and e_x respectively, and in all other degrees we put the zero module.

$$\begin{array}{ccccc} \cdots \to 0 \to & \mathfrak{K}_1(x) & \xrightarrow{d} & \mathfrak{K}_0(x) & \to 0 \to \cdots \\ & \| & & \| & \\ & A.e_x & & A.1 & \end{array}$$

The differential d in this chain complex is given by $d(a.e_x) = ax$ for $a \in A$.

Next, recall that if $\mathfrak{K}$ and $\mathfrak{K}'$ are chain complexes of A-modules, then the tensor product complex $\mathfrak{K} \otimes_A \mathfrak{K}'$ is defined by

$$(\mathfrak{K} \otimes_A \mathfrak{K}')_n = \bigoplus_{i+j=n} \mathfrak{K}_i \otimes_A \mathfrak{K}'_j.$$

If $x \in \mathfrak{K}_i$ and $y \in \mathfrak{K}'_j$, then

$$d(x \otimes y) = dx \otimes y + (-1)^i x \otimes dy.$$

The signs conspire so that $d \circ d = 0$ and the tensor product is again a chain complex.

If $x_1, \ldots, x_s$ are elements of A, we set

$$\mathfrak{K}(x_1, \ldots, x_s) = \mathfrak{K}(x_1) \otimes_A \cdots \otimes_A \mathfrak{K}(x_s).$$

Writing e_i for e_{x_i}, $\mathfrak{K}_r(x_1, \ldots, x_s)$ is a free A-module with one basis element $e_{i_1} \otimes \cdots \otimes e_{i_r}$ for each sequence $1 \leq i_1 < i_2 < \cdots < i_r \leq s$, and

$$d(a.e_{i_1} \otimes \cdots \otimes e_{i_r}) = \sum_{j=1}^{r} (-1)^{j+1} a.e_{i_1} \otimes \cdots \otimes e_{i_{j-1}} \otimes e_{i_{j+1}} \otimes \cdots \otimes e_{i_r}.$$

Remark Another (more general) way of constructing the Koszul complex is as follows. If M is an A-module and $f : M \to A$ is a map of A-modules, we write $\mathfrak{K}^A(f)$ for the complex

$$\cdots \xrightarrow{d} \Lambda^2 M \xrightarrow{d} \Lambda^1 M \xrightarrow{d} \Lambda^0 M$$

where the exterior powers are taken over A, and

$$d(m_1 \wedge \cdots \wedge m_r) = \sum_{j=1^r} (-1)^{j+1} f(m_j)(m_1 \wedge \cdots \wedge m_{j-1} \wedge m_{j+1} \wedge \cdots \wedge m_r).$$

The complex $\mathfrak{K}^A(x_1, \ldots, x_s)$ is then equal to $\mathfrak{K}^A(f)$ where $f : A^s \to A$ sends the basis elements to $x_1, \ldots, x_s$.

Lemma 4.2.1 *If $A = K[x_1, \ldots, x_s]$ is a polynomial ring, then $\mathfrak{K}^A(x_1, \ldots, x_s)$ is a free resolution of K as an A-module.*

Proof We prove by induction on r that $\mathfrak{K}^A(x_1, \ldots, x_r)$ is exact in positive degrees; the case $r = s$ then proves the lemma. For $r = 1$, the statement is clear, since x_1 is not a zero divisor, so we assume $r > 1$. We have a short exact sequence of chain complexes

$$0 \to \mathfrak{K}_n(x_1, \ldots, x_{r-1}) \otimes_A \mathfrak{K}_0(x_r) \to \mathfrak{K}_n(x_1, \ldots, x_r)$$
$$\to \mathfrak{K}_{n-1}(x_1, \ldots, x_{r-1}) \otimes_A \mathfrak{K}_1(x_r) \to 0$$

and hence a long exact sequence of homology groups

$$\cdots \to H_n(\mathfrak{K}(x_1, \ldots, x_{r-1})) \otimes_A \mathfrak{K}_1(x_r) \xrightarrow{\partial} H_n(\mathfrak{K}(x_1, \ldots, x_{r-1})) \otimes_A \mathfrak{K}_0(x_r)$$
$$\to H_n(\mathfrak{K}(x_1, \ldots, x_r)) \to H_{n-1}(\mathfrak{K}(x_1, \ldots, x_{r-1})) \otimes_A \mathfrak{K}_1(x_r) \xrightarrow{\partial} \cdots$$

It is easy to check that the switchback map ∂ is (up to sign) the map induced by the differential on $\mathfrak{K}(x_r)$, namely multiplication by x_r. Since $H_n(\mathfrak{K}(x_1, \ldots, x_{r-1})) = 0$ for $n > 0$ by induction, and for $n = 0$ the map ∂ is injective, it follows that $H_n(\mathfrak{K}(x_1, \ldots, x_r)) = 0$ for $n > 0$ as required. □

Theorem 4.2.2 *If $A = K[x_1, \ldots, x_r]$ is a graded polynomial ring, and M is a graded A-module, then the minimal resolution of M takes the form*

$$0 \to F_s \to F_{s-1} \to \cdots \to F_0 \to M \to 0.$$

Proof We prove this by using the fact that $\mathrm{Tor}_i^A(M, K)$ can be calculated by resolving in either the first or the second variable. See for example Hilton and Stammbach [46], Section IV.11.

First, we calculate $\mathrm{Tor}_i^A(M, K)$ using the minimal resolution. We have $F_i \otimes_A K = F_i/A^+F_i$, the vector space of generators of F_i. Since the resolution is minimal, the boundary map $F_i \otimes_A K \to F_{i-1} \otimes_A K$ is zero, and hence

$$\mathrm{Tor}_i^A(M, K) \cong F_i \otimes_A K \cong F_i/A^+F_i.$$

On the other hand, by Lemma 4.2.1, $\mathrm{Tor}_i^A(M, K)$ is the homology of the complex

$$M \otimes_A \mathfrak{K}_i^A(x_1, \ldots, x_s).$$

For $i > s$, this complex is zero, and hence $\mathrm{Tor}_i^A(M, K) = 0$. This implies that $F_i/A^+F_i = 0$ and hence $F_i = 0$. □

Corollary 4.2.3 (Hilbert's syzygy theorem)

If $y_1, \ldots, y_s$ are generators for $K[V]^G$ then $K[V]^G$ has a finite free resolution of length at most s as a module for $A = K[y_1, \ldots, y_s]$:

$$0 \to F_s \to \cdots \to F_0 \to K[V]^G \to 0.$$

The Poincaré series $p(K[V]^G, t)$ has the form

$$p(K[V]^G, t) = \sum_{i=0}^{s} (-1)^i p(F_i, t) = \frac{\sum_{i=0}^{s} (-1)^i p(F_i/A^+F_i, t)}{\prod_{i=1}^{s} (1 - t^{k_i})}$$

where $k_i = \deg(y_i)$. □

We define the **homological dimension** $\mathrm{hdim}_A(M)$ of a graded A-module M to be the minimal length of a projective resolution of M. This is equal to the largest value of t for which $\mathrm{Ext}_A^t(M, K) \neq 0$, and that it is also equal to the largest value of t for which $\mathrm{Tor}_t^A(M, K) \neq 0$. This is because the dimensions of these spaces are both equal to the number of free generators of the tth free module in the minimal resolution. The Hilbert syzygy theorem states that every graded A-module has homological dimension at most equal to s.

More generally, for any ring A, we say that A has **global dimension** s if s is the maximal value over all A-modules M of the homological dimension of M. Thus A has finite global dimension if and only if all A-modules have finite and bounded homological dimension. In this language, Hilbert's syzygy theorem says that a polynomial ring in s variables has global dimension s. Conversely, we shall see in Section 6.2 that if A is a commutative graded K-algebra, finitely generated by elements of positive degree, and having finite global dimension, then A is a polynomial ring over K.

4.3 Depth and Cohen–Macaulay rings

In this section, we prove a theorem of Hochster and Eagon [47], which states that as long as the characteristic of K is coprime to $|G|$, if V is a finite dimensional KG-module then the ring of invariants $K[V]^G$ is a Cohen–Macaulay ring. This means that there is a polynomial subring over which $K[V]^G$ is a finitely generated *free* module. According to Stanley [102], this theorem was part of the folklore of commutative algebra before the appearance of [47].

We begin with a brief discussion of depth and Cohen–Macaulay rings.

Definition 4.3.1 *If A is a commutative Noetherian ring and M is a finitely generated A-module, an element $a \in A$ is* **regular** *for M provided that $0 \neq M \neq aM$, and if $am = 0$ for $m \in M$ then $m = 0$ (i.e., a is not a zero divisor on M). A sequence $x_1, \dots, x_r \in A$ is a* **regular sequence** *for M if each x_i is regular for $M/(x_1M + \cdots + x_{i-1}M)$. Note that in particular this implies that $M/(x_1M + \cdots + x_rM) \neq 0$.*

The **depth** *of M is the length of the longest regular sequence for M. The depth of the ring A is its depth as an A-module, i.e., the length of the longest sequence $x_1, \dots, x_r$ with x_i a non zero-divisor in $A/(x_1, \dots, x_{i-1})$ and $A/(x_1, \dots, x_r) \neq 0$. The ring A or the module M is* **Cohen–Macaulay** *if its depth is equal to its Krull dimension.*

It is worth making a number of remarks about this definition. The first remark is that an element $a \in A$ with $0 \neq aM \neq M$ is regular for M if and only if it is not in any associated prime of M, by Lemma 2.2.1.

If A is local, then by the discussion of Section 2.3, a regular sequence $x_1, \dots, x_r$ for M can be extended to a system of parameters for M. In particular, by Theorem 2.3.2, the depth is at most equal to the Krull dimension of M. So if M is Cohen–Macaulay, then there is a regular sequence which is a system of parameters.

If $A = \bigoplus_{j=0}^{\infty} A_j$ is a commutative graded ring with $A_0 = K$ a field, and finitely generated over K by elements of positive degree, and M is a finitely generated graded A-module, then we demand that our regular sequences consist of homogeneous elements. We shall see in the next section, when we provide a homological characterization of depth, that the depth is independent of whether we make this demand. By Theorem 2.2.7, the depth is again at most equal to the Krull dimension of M, so that if M is Cohen–Macaulay, then there is a regular sequence which is a homogeneous system of parameters.

So we formulate the following hypothesis, which describes the generality in which we choose to work.

Hypothesis 4.3.2 *A is a commutative Noetherian ring, M is a finitely generated A-module, and either:*

(a) *A is local, with maximal ideal $\mathfrak{M}$, so that $K = A/\mathfrak{M}$ is a field, or*

(b) $A = \bigoplus_{j=0}^{\infty} A_j$ *and* $M = \bigoplus_{j=-\infty}^{\infty} M_j$ *are graded, with* $A_0 = K$ *a field, and* A *is finitely generated over* K *by elements of positive degree. In this case, we write* $\mathfrak{M}$ *for the ideal* A^+ *generated by the elements of positive degree.*

Lemma 4.3.3 *Under Hypothesis 4.3.2, the depth of* M *is at most equal to the Krull dimension of* M.

Proof This follows from the discussion preceding the hypothesis. □

Under this hypothesis, again examining the statement that an element $a \in A$ with $0 \neq aM \neq M$ is regular for M if and only if it is not in any associated prime of M, we see that if M is Cohen–Macaulay, of Krull dimension n, then for each associated prime $\mathfrak{p}$, $\dim(A/\mathfrak{p}) = n$. So an element $a \in A$ is not a zero divisor if and only if $\dim(M/aM) = \dim(M) - 1$; otherwise $\dim(M/aM) = \dim(M)$. So we have the following.

Proposition 4.3.4 (Macaulay) *Under Hypothesis 4.3.2, if* M *is Cohen–Macaulay then a sequence* $x_1, \ldots, x_r$ *is regular for* M *if and only if*

$$\dim(M/(x_1M + \cdots + x_rM)) = \dim(M) - r.$$

□

Theorem 4.3.5 *In case* (b) *of Hypothesis 4.3.2, the following are equivalent:*

(i) *M is Cohen–Macaulay.*

(ii) *There exist homogeneous elements $x_1, \ldots, x_n$ ($n = \dim(M)$) generating a polynomial subring $K[x_1, \ldots, x_n] \subseteq A/\mathrm{Ann}_A(M)$, such that M is a finitely generated free module over $K[x_1, \ldots, x_n]$.*

(iii) *Whenever $x_1, \ldots, x_n \in A$ are homogeneous elements generating a polynomial subring over which $K[x_1, \ldots, x_n] \subseteq A/\mathrm{Ann}_A(M)$ over which M is finitely generated, M is a free $K[x_1, \ldots, x_n]$-module.*

Proof By Noether Normalization (2.2.7), there is a polynomial subring generated by homogeneous elements $K[x_1, \ldots, x_n]$ over which $A/\mathrm{Ann}_A(M)$ and hence also M is finitely generated as a module, so that in particular $M/(x_1M + \cdots + x_nM)$ has Krull dimension zero. By the proposition, if M is Cohen–Macaulay, this implies that $x_1, \ldots, x_n$ is a regular sequence for M. Thus if $y_1, \ldots, y_t$ are homogeneous elements of M whose images form a vector space basis for $M/(x_1M + \cdots + x_nM)$, then $y_1, \ldots, y_t$ is a basis for M as a free module over $K[x_1, \ldots, x_n]$. This proves that (i) implies (iii) and (iii) implies (ii). To see that (ii) implies (i), if M is finitely generated and free over $K[x_1, \ldots, x_n]$ then $x_1, \ldots, x_n$ form a regular sequence for M. Since M has Krull dimension n, this implies that M is Cohen–Macaulay. □

Theorem 4.3.6 (Hochster–Eagon [47]) *Suppose that G is a finite group and K is a field of characteristic coprime to $|G|$. Then $K[V]^G$ is Cohen–Macaulay.*

Proof We may apply the Noether normalization theorem to $K[V]^G$ to obtain homogeneous elements $f_1, \ldots, f_n$ generating a polynomial subring $K[f_1, \ldots, f_n]$ over which $K[V]^G$ is finitely generated as a module. Since $K[V]$ is finitely generated as a module for $K[V]^G$, it is also finitely generated over $K[f_1, \ldots, f_n]$. Since $K[V]$ is a polynomial ring, it is Cohen–Macaulay, and so by the above theorem, $f_1, \ldots, f_n$ is a regular sequence in $K[V]$.

Now the map $\pi_G = \frac{1}{|G|}\sum_{g \in G} g : K[V] \to K[V]^G$ is a $K[V]^G$-module homomorphism (Lemma 1.6.2), and hence also a $K[f_1, \ldots, f_n]$-module homomorphism. So $K[V]^G$ is a direct summand of $K[V]$ as a $K[f_1, \ldots, f_n]$-module, and is hence free by Lemma 4.1.1. □

Even in the non-coprime case, we have the following.

Proposition 4.3.7 *Suppose that G is a finite group and K is a field. If V has dimension at least two, then $K[V]^G$ has depth at least two.*

Proof Choose a homogeneous element of positive degree $x \in K[V]^G$. Clearly x is not a zero divisor. We claim that the depth of $K[V]^G/(x)$ is at least one. If this is false, then there is an associated prime of $K[V]^G/(x)$ of dimension zero. Since all associated primes are homogeneous, this means the ideal of $K[V]^G/(x)$ consisting of all elements of positive degree is an associated prime. So there is a homogeneous element $u \in K[V]^G$ not in (x) such that the product of u with any invariant of positive degree is in (x).

Now x has only finitely many irreducible factors in $K[V]$ (up to multiplication by a unit), so we may choose w homogeneous of positive degree and coprime to x (recall that $K[V]$ is a unique factorization domain). Let y be the product of the images of w under the action of G, so that $y \in K[V]^G$ has no common factors with x. From the way we chose u, we have $uy = vx$ for some $v \in K[V]^G$. By unique factorization in $K[V]$, x divides u in $K[V]$, say $x = uz$. Again by uniqueness of factorization in $K[V]$, z must be G-invariant and so u is in the ideal generated by x in $K[V]^G$. This contradicts the way we chose u, and so $K[V]^G/(x)$ has depth at least one, which completes the proof of the proposition. □

Example (Bertin [16]; see also Almkvist and Fossum [5])
The invariants $\mathbf{F}_2[x_1, x_2, x_3, x_4]^{\mathbf{Z}/4}$ (where $\mathbf{Z}/4$ acts by cyclically permuting the x_i) have Krull dimension four and depth two. Note that by Corollary 3.9.3 this ring is a unique factorization domain. This provided the first known example of a unique factorization domain that was not Cohen–Macaulay.

4.4 Homological characterization of depth

In this section, we see the connection between the Hilbert syzygy theorem and the notion of depth. Namely, if $y_1, \ldots, y_s$ is a set of homogeneous generators of $K[V]^G$, then the length of a minimal resolution of $K[V]^G$ as a $K[y_1, \ldots, y_s]$-module (i.e., the homological dimension) plus the depth is equal to the number s of generators chosen. For this reason, we think of depth as "homological codimension". The results of this section are due to Serre [90].

As with the Hilbert syzygy theorem, this is really a general theorem about finitely generated graded $K[y_1, \ldots, y_s]$-modules, and so we work in this generality.

In order to understand this homological characterization of depth, we begin by asking when a regular sequence can be extended.

Proposition 4.4.1 *Suppose A is a commutative Noetherian ring and M is a finitely generated A-module. If I is an ideal in A, then the following are equivalent.*

(i) *There is no element $x \in I$ with* $\mathrm{Ann}_M(x) = \{0\}$.

(ii) $\mathrm{Ann}_M(I) \neq \{0\}$.

(iii) *I is contained in some associated prime $\mathfrak{p} \in \mathrm{Ass}_A(M)$.*

(iv) $\mathrm{Hom}_A(A/I, M) \neq 0$.

Proof There is no element $x \in I$ with $\mathrm{Ann}_M(x) = \{0\}$ if and only if I is contained in the union in A of all the annihilators of non-zero elements of M. Since the maximal annihilators are the associated primes, this happens if and only if I is contained in the union of the associated primes of M. By Lemma 1.4.3, this happens if and only if I is contained in some associated prime of M, in other words, if I is contained in the annihilator of some non-zero element of M. So this happens if and only if the annihilator in M of I is non-zero. This is equivalent to the existence of a non-zero A-module homomorphism from A/I to M, since such a homomorphism is determined by the image of the identity element, which must be some element annihilated by I. □

Proposition 4.4.2 *Suppose that $x_1, \ldots, x_n$ is a sequence of elements of an ideal I in A, which form a regular sequence for a finitely generated A-module M. Then*

$$\mathrm{Hom}_A(A/I, M/(x_1M + \cdots + x_nM)) \cong \mathrm{Ext}^n_A(A/I, M).$$

In particular, this sequence can be extended to a regular sequence of elements of I of length $n+1$ if and only if $\mathrm{Ext}^n_A(A/I, M) = 0$. This condition is independent of choice of sequence $x_1, \ldots, x_n$, so all sequences of maximal length have the same length, namely the smallest n for which $\mathrm{Ext}^n_A(A/I, M) \neq 0$.

Proof We prove this by induction on n. It is true for $n = 0$ if we interpret Ext^0_A as Hom_A. Now assume that $n \geq 1$. Since $x_2, \ldots, x_n$ is a regular sequence for M/x_1M, by induction we have

$$\mathrm{Hom}_A(A/I, M/(x_1M + \cdots + x_nM)) \cong \mathrm{Ext}^{n-1}_A(A/I, M/x_1M).$$

Look at the short exact sequence

$$0 \to M \xrightarrow{x_1} M \to M/x_1M \to 0.$$

This gives a long exact sequence

$$\mathrm{Ext}_A^{n-1}(A/I, M) \to \mathrm{Ext}_A^{n-1}(A/I, M/x_1M) \to \mathrm{Ext}_A^n(A/I, M) \xrightarrow{x_1} \mathrm{Ext}_A^n(A/I, M).$$

Now by induction the left hand term is $\mathrm{Hom}_A(A/I, M/(x_1, \dots, x_{n-1})M)$, which is zero by Proposition 4.4.1, since x_n exists. The action of x_1 on $\mathrm{Ext}_A^n(A/I, M)$ is zero since x_1 annihilates the A-module A/I, so the right hand map in the above sequence is zero. It follows that

$$\mathrm{Ext}_A^{n-1}(A/I, M/x_1M) \cong \mathrm{Ext}_A^n(A/I, M),$$

which completes the inductive proof. □

Theorem 4.4.3 *Under Hypothesis 4.3.2, the depth of M is equal to the* **homological codimension** $\mathrm{hcodim}_A(M)$*, which is defined to be the smallest $n \geq 0$ for which* $\mathrm{Ext}_A^n(K, M) \neq 0$.

Proof Apply the above proposition with $I = \mathfrak{M}$. Note that the condition

$$M/(x_1M + \cdots + x_nM) \neq 0$$

forces $x_1, \dots, x_n$ to lie in $\mathfrak{M}$ in both cases of the hypothesis. □

Theorem 4.4.4 *If M is a finitely generated graded module for a graded polynomial ring $A = K[y_1, \dots, y_s]$ (with each y_i in positive degree), then*

$$\mathrm{hdim}_A(M) + \mathrm{hcodim}_A(M) = s.$$

Proof If $\mathrm{hcodim}_A(M) = 0$ then there is an injective map $0 \to K \to M$. Since $\mathrm{Tor}_{s+1}^A = 0$, Tor_s^A is left exact, so we have

$$0 \to \mathrm{Tor}_s^A(K, K) \to \mathrm{Tor}_s^A(M, K).$$

Now the Koszul complex shows that $\mathrm{Tor}_s^A(K, K) \cong K$ and so $\mathrm{Tor}_s^A(M, K) \neq 0$. Thus $\mathrm{hdim}_A(M) = s$.

Now work by induction on $\mathrm{hcodim}_A(M)$, which we may assume is positive. So there exists a homogeneous element $x \in A$ of positive degree, which is regular for M,

$$0 \to M \xrightarrow{x} M \to M/xM \to 0$$

and $\mathrm{hcodim}_A(M/xM) = \mathrm{hcodim}_A(M) - 1$. We have a long exact sequence

$$\mathrm{Tor}_i^A(M, K) \xrightarrow{x} \mathrm{Tor}_i^A(M, K) \to \mathrm{Tor}_i^A(M/xM, K) \to \mathrm{Tor}_{i-1}^A(M, K) \xrightarrow{x} \mathrm{Tor}_{i-1}^A(M, K).$$

Since x annihilates the A-module K, it annihilates $\mathrm{Tor}^A_*(M,K)$, so this gives

$$0 \to \mathrm{Tor}^A_i(M,K) \to \mathrm{Tor}^A_i(M/xM,K) \to \mathrm{Tor}^A_{i-1}(M,K) \to 0$$

and hence $\mathrm{Tor}^A_i(M/xM,K) = 0$ if and only if $\mathrm{Tor}^A_{i-1}(M,K) = 0$. So

$$\mathrm{hdim}_A(M/xM) = \mathrm{hdim}_A(M) + 1.$$

This completes the proof by induction. □

Corollary 4.4.5 *If M is a Cohen–Macaulay graded module for $A = K[y_1,\dots,y_s]$ then*

$$\mathrm{hdim}_A(M) = s - \dim(M),$$

so that the minimal resolution stops after $s - \dim(M)$ terms. □

Corollary 4.4.6 *Suppose that K is a field of characteristic coprime to the order of G, and V is a KG-module of dimension n. If $K[V]^G$ has s generators then there are non-trivial syzygies of the tth kind if and only if $t \le s - n$.* □

4.5 The canonical module and Gorenstein rings

The following is a graded version of a theorem of Ischebeck [52].

Theorem 4.5.1 *Suppose that $A = K[y_1,\dots,y_s]$ is a graded polynomial ring, with the y_i in positive degree, and M and N are finitely generated graded A-modules. Then $\mathrm{Ext}^i_A(N,M) = 0$ for $i < \mathrm{depth}(M) - \dim(N)$.*

Proof We prove this by induction on the Krull dimension of N. If there is a non-zero homomorphism of A-modules $K \to N$, then we have a short exact sequence

$$0 \to K \to N \to N/K \to 0$$

and hence a long exact sequence

$$\cdots \to \mathrm{Ext}^{i-1}_A(K,M) \to \mathrm{Ext}^i_A(N/K,M) \to \mathrm{Ext}^i_A(N,M) \to \mathrm{Ext}^i_A(K,M) \to \cdots$$

We have $\mathrm{Ext}^i_A(K,M) = 0$ for $i < \mathrm{depth}(M)$ (Theorem 4.4.3), and so it suffices to prove the theorem with N replaced by N/K. Since N is Noetherian, after applying this finitely many times we end up with the situation where there are no non-zero homomorphisms $K \to N$. By Proposition 4.4.1 we deduce that there is an element x of positive degree in A which is a non zero-divisor for N. So we have a short exact sequence

$$0 \to N \xrightarrow{x} N \to N/xN \to 0$$

and N/xN has Krull dimension one less than that of N. This gives us a long exact sequence

$$\cdots \to \mathrm{Ext}^i_A(N/xN, M) \to \mathrm{Ext}^i_A(N, M) \xrightarrow{x} \mathrm{Ext}^i_A(N, M) \to \mathrm{Ext}^{i+1}_A(N/xN, M) \to \cdots$$

By the inductive hypothesis, $\mathrm{Ext}^i_A(N/xN, M) = 0$ for $i < \mathrm{depth}(M) - \dim(N) + 1$, and so for $i < \mathrm{depth}(M) - \dim(N)$ we deduce that multiplication by x is an isomorphism on $\mathrm{Ext}^i_A(N, M)$. This implies that $\mathrm{Ext}^i_A(N, M) = 0$, since x has positive degree. □

Corollary 4.5.2 *If M is a Cohen–Macaulay graded module of Krull dimension n for a graded polynomial ring $A = K[y_1, \ldots, y_s]$, with the y_i in positive degree, then* $\mathrm{Ext}^i_A(M, A) = 0$ *for $i \neq s - n$.*

Proof By Theorem 4.4.4, M has homological dimension $s-n$, so that $\mathrm{Ext}^i_A(M, A) = 0$ for $i > s - n$. By the above theorem we also have $\mathrm{Ext}^i_A(M, A) = 0$ for $i < s - n$. □

Proposition 4.5.3 *If $A = K[y_1, \ldots, y_s] \subseteq A' = K[y_1, \ldots, y_{s+1}]$ are graded polynomial rings with the y_i in positive degree, and M is a graded A'-module which is finitely generated and Cohen–Macaulay of dimension n as an A-module, then M is also Cohen–Macaulay as an A'-module, and there is an A-module isomorphism*

$$\mathrm{Ext}^{s+1-n}_{A'}(M, A') \cong \mathrm{Ext}^{s-n}_A(M, A).$$

Proof We have a short exact sequence of A'-modules

$$0 \to A' \otimes_A M \xrightarrow{1\otimes y_{s+1} - y_{s+1}\otimes 1} A' \otimes_A M \to M \to 0$$

and hence a long exact sequence

$$\cdots \to \mathrm{Ext}^{s-n}_{A'}(M, A') \to \mathrm{Ext}^{s-n}_{A'}(A' \otimes_A M, A') \xrightarrow{1\otimes y_{s+1} - y_{s+1}\otimes 1} \mathrm{Ext}^{s-n}_{A'}(A' \otimes_A M, A')$$
$$\to \mathrm{Ext}^{s+1-n}_{A'}(M, A') \to \mathrm{Ext}^{s+1-n}_{A'}(A' \otimes_A M, A') \to \cdots$$

By the above corollary, the left hand term is zero.

Now if

$$\cdots \to F_1 \to F_0 \to M \to 0$$

is a free resolution of M as an A-module then

$$\cdots \to A' \otimes_A F_1 \to A' \otimes_A F_0 \to A' \otimes_A M \to 0$$

is a resolution of $A' \otimes_A M$ as an A'-module (it is exact since A' is a free A-module) and

$$\mathrm{Hom}_{A'}(A' \otimes_A F_i, A') \cong \mathrm{Hom}_A(F_i, A') \cong A' \otimes_A \mathrm{Hom}_A(F_i, A)$$

as A'-modules. So

$$\mathrm{Ext}^i_{A'}(A' \otimes_A M, A') \cong \mathrm{Ext}^i_A(F_i, A') \cong A' \otimes_A \mathrm{Ext}^i_A(M, A).$$

It follows that the right hand term in the above long exact sequence is zero, and so $\mathrm{Ext}^{s+1-n}_{A'}(M, A')$ is isomorphic to the cokernel of $1 \otimes y_{s+1} - y_{s+1} \otimes 1$ on $A' \otimes_A \mathrm{Ext}^{s-n}_A(M, A)$, which as an A-module is isomorphic to $\mathrm{Ext}^{s-n}_A(M, A)$. □

Remark Since A, A' and M are graded, we should really keep track of the grading. The above proof identifies $\mathrm{Ext}^{s+1-n}_{A'}(M, A')$ with $\mathrm{Ext}^{s-n}_A(M, A)$ shifted in degree by $\deg(y_{s+1})$. We write this as follows:

$$\mathrm{Ext}^{s+1-n}_{A'}(M, A') \cong \mathrm{Ext}^{s-n}_A(M, A)[\deg(y_{s+1})].$$

Now if R is a commutative Cohen–Macaulay graded ring of Krull dimension n, which is finitely generated over a field K by homogeneous elements $y_1, \dots, y_s$ of positive degree, then by Proposition 4.5.3, R is also a graded Cohen–Macaulay module for $A = K[y_1, \dots, y_s]$, and we define the **canonical module** $\omega_A(R)$ [42] to be

$$\mathrm{Ext}^{s-n}_A(R, A)[(\textstyle\sum_i \deg(y_i)) - n].$$

The kernel of $A \to R$ acts as zero on the left hand variable of this Ext group, so it also acts as zero on the A-module $\omega_A(R)$, so we may regard it as an R-module. By Proposition 4.5.3 and 4.3.5, if we add in a redundant generator y_{s+1}, we do not alter the canonical module (up to isomorphism). Since we may pass from any set of generators to any other by adding and taking away generators in this way, the canonical module is independent of choice of A, and so we simply write $\omega(R)$.

In fact, the $y_1, \dots, y_s$ in the above definition do not have to generate R. They only have to generate a subring over which R is finitely generated as a module. By Theorem 4.3.5, we may choose such elements with $s = n$, generating a polynomial ring over which R is finitely generated as a module. So we have the following:

Proposition 4.5.4 *Suppose that R is a commutative Cohen–Macaulay graded ring, finitely generated over a field K by elements of positive degree. Let $B = K[f_1, \dots, f_n]$ be a polynomial subring over which R is finitely generated as a module. Then as R-modules, we have*

$$\omega(R) \cong \mathrm{Hom}_B(R, B)[\textstyle\sum_i(\deg(f_i) - 1)].$$

□

If we had made this as our original definition, it would have been unclear that it was independent of the choice of $f_1, \dots, f_n$.

If $y_1, \dots, y_s$ are generators for R, then by Corollary 4.4.5, the minimal resolution of R as a module for $A = K[y_1, \dots, y_s]$ takes the form

$$0 \to F_{s-n} \to \cdots \to F_0 \to R \to 0.$$

We then set $F_i^* = \mathrm{Hom}_A(F_i, A)$, the dual free module. By Corollary 4.5.2, the dual of the above sequence of free modules is exact except at the end, and the homology there is by definition $\omega(R)$. So we obtain an exact sequence

$$0 \to F_0^* \to F_1^* \to \cdots \to F_{s-n}^* \to \omega(R) \to 0.$$

This must be the minimal resolution of $\omega(R)$ as an A-module, since otherwise we could dualize again and obtain a smaller resolution of R. If M is an A-module, then

$$\begin{aligned} \mathrm{Hom}_A(F_i^*, M) &\cong F_i \otimes_A M \\ \mathrm{Hom}_A(F_i, M) &\cong F_i^* \otimes_A M, \end{aligned}$$

and so we have

$$\begin{aligned} \mathrm{Ext}_A^i(\omega(R), M) &\cong \mathrm{Tor}_{s-n-i}^A(M, R) \\ \mathrm{Tor}_i^A(\omega(R), M) &\cong \mathrm{Ext}_A^{s-n-i}(M, R). \end{aligned}$$

Definition 4.5.5 *If R is a commutative Cohen–Macaulay graded ring, finitely generated over K by elements of positive degree, we say that R is* **Gorenstein** *if $\omega(R) \cong R$, possibly with a shift in degree. If $\omega(R) \cong R$ with no shift in degree, we say that R is* **graded Gorenstein.**

Thus for example if R is a polynomial ring $K[x_1, \dots, x_n]$ then we may take $B = R$, and so

$$\omega(R) = R[\textstyle\sum_i(\deg(x_i) - 1)].$$

So R is Gorenstein, but it is only graded Gorenstein if the polynomial generators lie in degree one.

Let us examine in more detail the case where the generators lie in degree one. Suppose, instead of taking $B = R$, we take $B = K[f_1, \dots, f_n]$, where the f_i are polynomials of degree k_i generating a polynomial subring over which R is finitely generated as a module. Then a basis for R as a free B-module may be found by choosing a vector space basis for $R/(f_1, \dots, f_n)$ and lifting back to R. The Poincaré series is given by

$$p(R/(f_1, \dots, f_n)) = \frac{\prod_i(1 - t^{k_i})}{(1-t)^n} = \prod_i (1 + t + \cdots + t^{k_i - 1}).$$

So $R/(f_1, \dots, f_n)$ is one dimensional in degree $\sum_i(k_i - 1)$. Choose a basis element α in this degree (this is only well defined up to scalar multiplication). We have seen in Proposition 4.5.4 that R is isomorphic to $\mathrm{Hom}_B(R, B)[\sum_i(k_i - 1)]$ as graded R-modules. We can identify such an isomorphism by seeing what homomorphism corresponds to the identity element of R. By the above dimension count, the identity element goes to a multiple of the map α^* sending α to 1 and elements of degree less than $\sum_i(k_i - 1)$ to zero.

4.6 Watanabe's theorem

In this section, we identify the canonical module $\omega(K[V]^G)$ in the case where the characteristic of K is coprime to the order of G. This is implicit in the work of Watanabe [107, 108], and explicit in unpublished work of Eisenbud, according to Stanley [102].

Let V be a finite dimensional representation of a finite group G over a field K whose characteristic is coprime to $|G|$. We examine what the last section tells us in case $R = K[V]$. Choose a polynomial subring $B = K[f_1, \dots, f_n]$ generated by homogeneous elements f_i of degree k_i, over which $K[V]^G$ is finitely generated as a module. Since $K[V]$ is finitely generated as a module over $K[V]^G$, it is also finitely generated as a module over B. So we may apply the discussion at the end of the last section. In order to understand the action of G on $\mathrm{Hom}_B(K[V], B)$, it suffices to understand the action on $K[V]/(f_1, \dots, f_n)$. Assume for convenience that the exponent of G divides each k_i, so that if $g \in G$ has eigenvalues $\lambda_1, \dots, \lambda_n$ on V, then $\lambda_i^{k_i} = 1$. Using Proposition 2.5.1, we have

$$\begin{aligned}\sum_{j=0}^{\infty} t^j \mathrm{Tr}(g, (K[V]/(f_1, \dots, f_n))_j) &= \prod_{i=1}^{n}(1 - t^{k_i}) \sum_{j=0}^{\infty} t^j \mathrm{Tr}(g, K[V]_j) \\ &= \frac{\prod_{i=1}^{n}(1 - t^{k_i})}{\det(1 - g^{-1}t, V)} = \prod_{i=1}^{n} \left(\frac{1 - t^{k_i}}{1 - \lambda_i^{-1} t} \right) \\ &= \prod_{i=1}^{n} (1 + \lambda_i^{-1} t + \lambda_i^{-2} t^2 + \dots + \lambda t^{k_i - 1}).\end{aligned}$$

Recall that $K[V]/(f_1, \dots, f_n)$ is one dimensional in degree $\sum_i(k_i - 1)$, and that we have chosen a basis element α in this degree. The above calculation shows that the action of g on this one dimensional space is equal to multiplication by $\prod_{i=1}^{n} \lambda_i = \det(g, V)$. It follows that if we expect the isomorphism

$$K[V] \cong \mathrm{Hom}_B(K[V], B)[\textstyle\sum_i(k_i - 1)]$$

to commute with the G-action, we should really tensor $K[V]$ with the inverse of the determinant representation first.

Proposition 4.6.1 *We have an isomorphism of $K[V]$ modules, commuting with the G-action*

$$K[V] \otimes \det{}^{-1} \cong \mathrm{Hom}_B(K[V], B)[\textstyle\sum_i(k_i - 1)].$$

Proof Denote by ξ the homomorphism from the left hand side to the right hand side described at the end of the last section. In other words, if $f \in K[V]$ and $a \in K$, then

$$\xi(f \otimes a)(x) = a.\alpha^*(f.x).$$

We already know that this is an isomorphism of $K[V]$-modules, so it suffices to check that it commutes with the G-action. We showed above that $g(\alpha) = \det(g, V)\alpha$, and so $g(\alpha^*) = \det(g^{-1}, V)\alpha^*$. So we have

$$\begin{aligned} g(\xi(f \otimes a))(x) &= (\xi(f \otimes a))(g^{-1}(x)) = a.\alpha^*(f.g^{-1}(x)) \\ &= a.g(\alpha^*)(f.x) = a.\det(g^{-1}, V)\alpha^*(f.x) \\ &= \xi(f \otimes a.\det(g^{-1}, V))(x) = \xi(g(f \otimes a))(x). \end{aligned}$$

□

Theorem 4.6.2 (Watanabe) *Suppose that V is a finite dimensional representation of a finite group G over a field K of characteristic coprime to $|G|$. Then the canonical module for $K[V]^G$ is given by*

$$\omega(K[V]^G) \cong K[V]^G_{\det}.$$

In particular, $K[V]^G$ is graded Gorenstein if and only if $G \subseteq SL(V)$.

Proof Let B be as above. We have

$$\begin{aligned} \omega(K[V]^G) &\cong \mathrm{Hom}_B(K[V]^G, B)[\textstyle\sum_i(k_i - 1)] \\ &\cong (\mathrm{Hom}_B(K[V], B)[\textstyle\sum_i(k_i - 1)])^G \\ &\cong (K[V] \otimes \det{}^{-1})^G \cong K[V]^G_{\det}. \end{aligned}$$

□

Chapter 5

Polynomial tensor exterior algebras

5.1 Motivation and first properties

In this chapter, we examine the invariants of G on the tensor product $K[V] \otimes \Lambda(V^*)$ of the symmetric algebra on V^* with the exterior algebra on V^*. If $K[V] = K[x_1, \dots, x_n]$ then $K[V] \otimes \Lambda(V^*)$ can be thought of as the Grassmann algebra of differential forms

$$K[x_1, \dots, x_n] \otimes \Lambda(dx_1, \dots, dx_n)$$

so that a typical element is a linear combination of terms of the form $f\, dx_{i_1} \wedge \dots \wedge dx_{i_j}$ with $\{i_1, \dots, i_j\} \subseteq \{1, \dots, n\}$. Then the elements of $(K[V] \otimes \Lambda(V^*))^G$ are the G-invariant differential forms on V. It is customary to grade the algebra of differential forms in such a way that the exterior generators are in degree one, and the polynomial generators are in degree two. This means that it is **graded commutative** in the sense that if $\deg(x) = a$ and $\deg(y) = b$ then $xy = (-1)^{ab} yx$.

We have a derivation d of degree -1 on $K[V] \otimes \Lambda(V^*)$ given by sending x_i to dx_i and dx_i to zero. Thus $f\, dx_{i_1} \wedge \dots \wedge dx_{i_j}$ is sent to $df \wedge dx_{i_1} \wedge \dots \wedge dx_{i_j}$. Note that d commutes with the G-action, so that it sends invariant elements to invariant elements.

One of the motivations for considering the invariant theory of polynomial tensor exterior algebras is as follows. Suppose that E is an elementary abelian p-group $(\mathbb{Z}/p)^n$, and G is a group of order coprime to p acting on E by group automorphisms. If K is a field of characteristic p then we have

$$H^*(E, K) = \mathrm{Ext}^*_{KE}(K, K) = \begin{cases} K[V] & p = 2 \\ K[V] \otimes \Lambda(V^*) & p > 2 \end{cases}$$

where $V = K \otimes_{\mathbb{F}_p} E$. If E is a Sylow p-subgroup of a finite group H, with $G = N_H(E)/E$, then

$$H^*(H,K) \cong H^*(E,K)^G = \begin{cases} K[V]^G & p=2 \\ (K[V] \otimes \Lambda(V^*))^G & p>2. \end{cases}$$

5.2 A variation on Molien's theorem

In this section, we prove the analogue of Molien's theorem, giving the Poincaré series of the polynomial tensor exterior algebra $(K[V] \otimes \Lambda(V^*))^G$. We begin with the analogue of Proposition 2.5.1.

Proposition 5.2.1 $\sum_{j=0}^{\infty} t^j \mathrm{Tr}(g,(K[V] \otimes \Lambda(V^*))_j) = \dfrac{\det(1+g^{-1}t,V)}{\det(1-g^{-1}t^2,V)}.$

As in Proposition 2.5.1, we may assume that K is algebraically closed, and that the action of g on V is upper triangular with eigenvalues $\lambda_1, \dots, \lambda_n$. The eigenvalues on $\Lambda^j(V^*)$ are the products of j distinct λ_i^{-1}'s. So we have

$$\sum_{j=0}^{\infty} t^j \mathrm{Tr}(g, \Lambda^j(V^*)) = \prod_{i=1}^{n}(1+\lambda_i^{-1}t) = \det(1+g^{-1}t,V).$$

Recall that we are putting the polynomial generators in degree two, so that

$$\sum_{j=0}^{\infty} t^j \mathrm{Tr}(g, K[V]_j) = \frac{1}{\det(1-g^{-1}t^2,V)}.$$

To complete the proof, we observe that

$$\sum_{j=0}^{\infty} t^j \mathrm{Tr}(g,(K[V] \otimes \Lambda(V^*))_j) = \left(\sum_{j=0}^{\infty} t^j \mathrm{Tr}(g,K[V]_j)\right)\left(\sum_{j=0}^{\infty} t^j \mathrm{Tr}(g,\Lambda^j(V^*))\right).$$

□

Theorem 5.2.2 *If K is a field of characteristic zero then*

$$p((K[V] \otimes \Lambda(V^*))^G, t) = \frac{1}{|G|}\sum_{g \in G} \frac{\det(1+g^{-1}t,V)}{\det(1-g^{-1}t^2,V)}.$$

Proof As in Section 2.5, this follows from the above proposition, using the fact that the dimension of the space of fixed points is equal to the trace of the projection $\pi_G = \frac{1}{|G|}\sum_{g \in G} g$. □

5.3 The invariants are graded Gorenstein

In this section, we prove the analogue of Watanabe's theorem (Section 4.6) for a polynomial tensor exterior algebra. In contrast with the case of a polynomial ring, the invariants are always graded Gorenstein.

First, we must discuss the fact that $K[V] \otimes \Lambda(V^*)$ is not commutative, so that at first sight, the usual theorems of commutative algebra do not apply. However, it is graded commutative, in the sense that if $\deg(x) = a$ and $\deg(y) = b$ then $xy = (-1)^{ab}yx$. One can easily check that all the theorems we have written out for strictly commutative graded rings hold with exactly the same proof for graded commutative rings in the above sense.

The following is the analogue of the theorem of Hochster and Eagon (4.3.6).

Proposition 5.3.1 *If V is a finite dimensional representation of a finite group G over a field K whose characteristic does not divide $|G|$, then $(K[V] \otimes \Lambda(V^*))^G$ is a Cohen–Macaulay graded commutative ring.*

Proof As in the proof of Theorem 4.3.6, we choose a set of homogeneous elements $f_1, \dots, f_n$ in $K[V]^G$ generating a polynomial ring over which $K[V]$ is finitely generated as a module. Since $K[V] \otimes \Lambda(V^*)$ is a finitely generated free module over $K[V]$, it is also a finitely generated free module over $K[f_1, \dots, f_n]$. The rest of the proof is the same as in Theorem 4.3.6. □

Theorem 5.3.2 *If V is a finite dimensional representation of a finite group G over a field K whose characteristic does not divide $|G|$, then $(K[V] \otimes \Lambda(V^*))^G$ is graded Gorenstein.*

Proof As in Section 4.6, we choose a polynomial subring $B = K[f_1, \dots, f_n] \subseteq K[V]^G$ generated by homogeneous elements f_i of degree $2k_i$ (remember the generators of $K[V]$ are now in degree two), over which $K[V]$ and hence also $K[V] \otimes \Lambda(V^*)$ is finitely generated as a module. Again assume that the exponent of G divides each k_i. If $g \in G$ has eigenvalues $\lambda_1, \dots, \lambda_n$ on V, then using Proposition 5.2.1, we have

$$\begin{aligned}
&\sum_{j=0}^{\infty} t^j \operatorname{Tr}(g, (K[V] \otimes \Lambda(V^*)/(f_1, \dots, f_n))_j) \\
&= \prod_{i=1}^{n}(1 - t^{2k_i}) \sum_{j=0}^{\infty} t^j \operatorname{Tr}(g, (K[V] \otimes \Lambda(V^*))_j) \\
&= \frac{\prod_{i=1}^{n}(1 - t^{2k_i}) \det(1 + g^{-1}t, V)}{\det(1 - g^{-1}t^2, V)} = \prod_{i=1}^{n} \frac{(1 - t^{2k_i})(1 + \lambda^{-1}t)}{1 - \lambda^{-1}t^2} \\
&= \prod_{i=1}^{n}(1 + \lambda_i^{-1}t + \lambda_i^{-1}t^2 + \dots + t^{2k_i - 1}).
\end{aligned}$$

Now $K[V] \otimes \Lambda(V^*)$ is graded Gorenstein, and $(K[V] \otimes \Lambda(V^*))/(f_1, \dots, f_n)$ is one dimensional in degree $\sum_i (2k_i - 1)$. Choosing a basis element α in this degree, the above calculation shows that G acts trivially on α. Exactly the same calculation as given in Proposition 4.6.1 now shows that we have an isomorphism

$$K[V] \otimes \Lambda(V^*) \cong \mathrm{Hom}_B(K[V] \otimes \Lambda(V^*), B)[\textstyle\sum_i (2k_i - 1)]$$

of $K[V] \otimes \Lambda(V^*)$-modules, commuting with the G-action. Thus

$$\begin{aligned} \omega\left((K[V] \otimes \Lambda(V^*))^G\right) &\cong \mathrm{Hom}_B((K[V] \otimes \Lambda(V^*))^G, B)[\textstyle\sum_i (2k_i - 1)] \\ &\cong (\mathrm{Hom}_B(K[V] \otimes \Lambda(V^*), B)[\textstyle\sum_i (2k_i - 1)])^G \\ &\cong (K[V] \otimes \Lambda(V^*))^G \end{aligned}$$

and so $(K[V] \otimes \Lambda(V^*))^G$ is graded Gorenstein. □

Remark Suppose that K is a field of characteristic p and H is a finite group with elementary abelian Sylow p-subgroup E. Set $G = N_H(E)/E$ and $V = K \otimes_{\mathbb{F}_p} E$ (see Section 5.1). If p is odd, then by the above theorem $H^*(H, K) = (K[V] \otimes \Lambda(V^*))^G$ is graded Gorenstein. If $p = 2$ then G acts on E as special linear transformations, because $\mathbb{F}_2$ has only one non-zero element, and so by Theorem 4.6.2, $H^*(H, K) = K[V]^G$ is again graded Gorenstein. These are special cases of a theorem (Benson and Carlson [13]) which states that if H is any finite group and $H^*(H, K)$ is Cohen–Macaulay, then it is graded Gorenstein. The proof does not involve treating the elementary abelian case first.

5.4 The Jacobian

Now suppose that $K[f_1, \dots, f_n] \subseteq K[V]^G$ is a polynomial subring generated by homogeneous elements, over which $K[V]^G$ is finitely generated as a module (Noether Normalization 2.2.7). Regard $\partial f_i / \partial x_j$ as an $n \times n$ matrix, and let $\mathfrak{j} = \det(\partial f_i / \partial x_j)$ be its determinant, the **Jacobian** of the f_i with respect to the x_j. Then

$$df_1 \wedge \dots \wedge df_n = \mathfrak{j}\, dx_1 \wedge \dots \wedge dx_n$$

is G-invariant. Since

$$g(dx_1 \wedge \dots \wedge dx_n) = \det(g^{-1}, V)\, dx_1 \wedge \dots \wedge dx_n,$$

we have

$$g(\mathfrak{j}) = \det(g, V)\, \mathfrak{j}$$

so that $\mathfrak{j}$ is a relative invariant for the determinant representation. The following proposition gives us a criterion for the non-vanishing of $\mathfrak{j}$. This will be used later, in Sections 7.3 and 8.3. We begin with a lemma.

Lemma 5.4.1 *Suppose that K is a field, and that K' is a finite separable extension of K. Suppose that $x_1, \ldots, x_n$ are elements of K' with minimal equations $h_1(X_1), \ldots, h_n(X_n)$ respectively. If $\xi \in K[X_1, \ldots, X_n]$ is any polynomial such that $\xi(x_1, \ldots, x_n) = 0$ in K', then there is an element $\rho(X_1, \ldots, X_n) \in K[X_1, \ldots, X_n]$ with $\rho(x_1, \ldots, x_n) \neq 0$, such that $\rho(X_1, \ldots, X_n)\xi(X_1, \ldots, X_n)$ is in the ideal generated by $h_1(X_1), \ldots, h_n(X_n)$.*

Proof This is proved by induction on n. The case $n = 1$ is clear, so suppose that the lemma is true for $n - 1$. Then K' is still a finite separable extension of $K(x_n)$. Let $h'_i(X_i, x_n)$ be the minimal equation of x_i over $K(x_n)$ for $1 \leq i \leq n - 1$. Then $h'_i(X_i, x_n)$ appears exactly once in the factorization of $h_i(X_i)$ into irreducible factors over $K(x_n)$ by separability, so we have $h_i(X_i) = h'_i(X_i, x_n)h''_i(X_i, x_n)$ with $h''_i(X_i, x_n) \neq 0$. By the inductive hypothesis there is a non-zero element $\rho'(X_1, \ldots, X_{n-1}, x_n) \in K(x_n)[X_1, \ldots, X_{n-1}]$ with $\rho'(x_1, \ldots, x_{n-1}, x_n) \neq 0$, such that $\rho'(X_1, \ldots, X_{n-1}, x_n)\xi(X_1, \ldots, X_{n-1}, x_n)$ is in the ideal generated by the elements $h'_1(X_1, x_n), \ldots, h'_{n-1}(X_{n-1}, x_n)$. Setting

$$\rho(X_1, \ldots, X_n) = h''_1(X_1, X_n) \cdots h''_{n-1}(X_{n-1}, X_n)\rho'(X_1, \ldots, X_{n-1}, X_n)$$

we see that $\rho(x_1, \ldots, x_n) \neq 0$. Furthermore, $\rho(X_1, \ldots, X_{n-1}, x_n)\xi(X_1, \ldots, X_{n-1}, x_n)$ is in the ideal of $K[X_1, \ldots, X_n]$ generated by $h_1(X_1), \ldots, h_{n-1}(X_{n-1})$, and so the element $\rho(X_1, \ldots, X_n)\xi(X_1, \ldots, X_n)$ is in the ideal generated by $h_1(X_1), \ldots, h_n(X_n)$.
□

Proposition 5.4.2 *Let $x_1, \ldots, x_n$ be algebraically independent indeterminates over a perfect field K. If $f_1, \ldots, f_n$ are elements of $K(x_1, \ldots, x_n)$, then $K(x_1, \ldots, x_n)$ is a finite separable extension of $K(f_1, \ldots, f_n)$ if and only if $\mathrm{j} = \det(\partial f_i/\partial x_j) \neq 0$.*

Proof First, suppose that $K(x_1, \ldots, x_n)$ is an infinite extension of $K(f_1, \ldots, f_n)$. Then some x_j is not algebraic over $K(f_1, \ldots, f_n)$. If $f_1, \ldots, f_n$ were algebraically independent, then the subfield of $K(x_1, \ldots, x_n)$ generated over K by $x_j, f_1, \ldots, f_n$ would have transcendence degree $n + 1$. This is impossible, and so $f_1, \ldots, f_n$ are algebraically dependent over K. So there is a non-zero polynomial $\phi \in K[F_1, \ldots, F_n]$ with $\phi(f_1, \ldots, f_n) = 0$. Choose such a ϕ of smallest possible degree. For convenience, for the rest of this proof we use a bar to indicate that small letters have been substituted for the corresponding large ones. So for example this equation reads $\bar{\phi} = 0$. Differentiating, we get

$$0 = \frac{\partial \bar{\phi}}{\partial x_j} = \sum_i \frac{\partial \bar{\phi}}{\partial f_i} \frac{\partial f_i}{\partial x_j}.$$

Here, $\partial \bar{\phi}/\partial f_i$ denotes $\partial \phi(F_1, \ldots, F_n)/\partial F_i$ with the functions $f_1, \ldots, f_n$ substituted for $F_1, \ldots, F_n$. If all the $\partial \phi/\partial F_i$ are zero, then since K is perfect, we can write

$\phi = (\phi_0)^p$. But then $\bar{\phi}_0 = 0$, so by minimality of the degree of ϕ, this cannot happen. This means that at least one of the $\partial\phi/\partial F_i$ is a non-zero polynomial. It has smaller degree than ϕ, and so $\partial\bar{\phi}/\partial f_i$ is non-zero. It follows that the above equation gives a non-trivial linear relation between the rows of the matrix $\partial f_i/\partial x_j$, and so $\mathrm{j} = \det(\partial f_i/\partial x_j) = 0$.

Next, suppose that $K(x_1, \dots, x_n)$ is a finite extension of $K(f_1, \dots, f_n)$, but not separable. Then some x_j is inseparable over $K(f_1, \dots, f_n)$. The minimal equation of this x_j is therefore of the form

$$x_j^{mp} + \bar{\psi}_{m-1} x_j^{(m-1)p} + \cdots + \bar{\psi}_0 = 0$$

for some $\psi_j \in K[F_1, \dots, F_n]$. Differentiate this equation with respect to x_j to get

$$\sum_i \left(\frac{\partial \bar{\psi}_{m-1}}{\partial f_i} x_j^{(m-1)p} + \cdots + \frac{\partial \bar{\psi}_0}{\partial f_i} \right) \frac{\partial f_i}{\partial x_j} = 0.$$

The coefficients of the $\partial f_i/\partial x_j$ in this equation have smaller degree in x_j than the degree of the minimal equation, so they can only all vanish if all $\partial\bar{\psi}_j/\partial f_i$ vanish. Since $f_1, \dots, f_n$ are algebraically independent, this would mean that each ψ_j is a pth power (since K is perfect). But the minimal equation of x_j cannot be a pth power, so the above equation gives us a non-trivial linear relation between the rows of $\partial f_i/\partial x_j$, and so $\mathrm{j} = \det(\partial f_i/\partial x_j) = 0$.

Finally, suppose that $K(x_1, \dots, x_n)$ is a finite separable extension of the field $K(f_1, \dots, f_n)$. We prove that $\det(\partial f_i/\partial x_j) \neq 0$ by finding an inverse by a process known as "algebraic differentiation". This involves introducing rational functions $\partial x_i/\partial f_j$ with the property that the matrix $(\partial x_i/\partial f_j)$ is inverse to the matrix $(\partial f_i/\partial x_j)$.

We first give a heuristic justification for the definition we shall use for $\partial x_i/\partial f_j$. If z is any element of $K(x_1, \dots, x_n)$, then it is algebraic over $K(f_1, \dots, f_n)$, say with minimal equation

$$h(f_1, \dots, f_n, Z) = Z^s + a_{s-1} Z^{s-1} + \cdots + a_0$$

with $a_j \in K(f_1, \dots, f_n)$, and separability implies that $\partial h/\partial Z \neq 0$. Regarding $h(z) = 0$ as an identity between functions, we differentiate with respect to f_j to get

$$s x_i^{s-1} \frac{\partial z}{\partial f_j} + \frac{\partial a_{s-1}}{\partial f_j} z^{s-1} + (s-1) a_{s-1} z^{s-2} \frac{\partial z}{\partial f_j} + \cdots + \frac{\partial a_0}{\partial f_j} = 0.$$

In other words,

$$\frac{\partial h}{\partial z} \frac{\partial z}{\partial f_j} + \frac{\partial h}{\partial f_j} = 0,$$

where $\partial h/\partial z$ denotes $\partial h/\partial Z$ with z substituted for Z. Since $\partial h/\partial z \neq 0$, this means that

$$\frac{\partial z}{\partial f_j} = \frac{-\partial h/\partial f_j}{\partial h/\partial z}.$$

The above argument suggests that we use this formula as the *definition* of $\partial z/\partial f_j$.

Note that if also $H(z) = 0$ then $H(Z) = h(Z)\phi(Z)$ for some polynomial $\phi(Z)$, and so

$$\frac{\partial H}{\partial z}\frac{\partial z}{\partial f_j} + \frac{\partial H}{\partial f_j} = h(z)\left(\frac{\partial \phi}{\partial z}\frac{\partial z}{\partial f_j} + \frac{\partial \phi}{\partial f_j}\right) + \phi(z)\left(\frac{\partial h}{\partial z}\frac{\partial z}{\partial f_j} + \frac{\partial h}{\partial f_j}\right) = 0.$$

We claim that for any $z \in K(x_1, \ldots, x_n)$ we have

$$\frac{\partial z}{\partial f_j} = \sum_i \frac{\partial z}{\partial x_i}\frac{\partial x_i}{\partial f_j}.$$

In the polynomial ring $K[X_1, \ldots, X_n, F_1, \ldots, F_n, Z]$ in $2n+1$ indeterminates, we let $h(F_1, \ldots, F_n, Z)$, $h_i(F_1, \ldots, F_n, X_i)$ and $\xi(X_1, \ldots, X_n, Z)$ be polynomials which, when the appropriate small letters are substituted for large letters, give the minimal equations of z and x_i over $K(f_1, \ldots, f_n)$ and z over $K(x_1, \ldots, x_n)$ respectively, with the bottoms of the fractions cleared by multiplying out in the obvious way. By the lemma, there is a polynomial $\rho(X_1, \ldots, X_n, F_1, \ldots, F_n, Z)$ with $\bar{\rho} \neq 0$, and such that $\rho\xi$ lies in the ideal generated by h and the h_i:

$$\rho\xi = \eta h + \sum_i \eta_i h_i$$

for suitable polynomials η and η_i. Differentiate with respect to each of the $2n+1$ variables in turn, and substitute small letters for the large ones. We obtain

$$\begin{aligned} \bar{\rho}\frac{\partial \bar{\xi}}{\partial x_j} &= \bar{\eta}_j \frac{\partial \bar{h}_j}{\partial x_j} \\ 0 &= \bar{\eta}\frac{\partial \bar{h}}{\partial f_j} + \sum_i \eta_i \frac{\partial \bar{h}_i}{\partial f_j} \\ \bar{\rho}\frac{\partial \bar{\xi}}{\partial z} &= \bar{\eta}\frac{\partial \bar{h}}{\partial z}. \end{aligned}$$

Thus we have

$$\frac{\partial z}{\partial f_j} = -\frac{\eta \partial \bar{h}/\partial f_j}{\eta \partial \bar{h}/\partial z} = \sum_i \frac{\eta_i \partial \bar{h}_i/\partial f_j}{\rho \partial \bar{\xi}/\partial z} = \sum_i \frac{\partial \bar{xi}/\partial x_i}{\partial \bar{\xi}/\partial z}\frac{\partial \bar{h}_i/\partial f_j}{\partial \bar{h}_i/\partial x_i} = \sum_i \frac{\partial z}{\partial x_i}\frac{\partial x_i}{\partial f_j}.$$

In particular, putting $z = f_j$ we see that $\sum_i(\partial f_j/\partial x_i)(\partial x_i/\partial f_j)$ takes the value one if $i = j$ and zero otherwise. So the matrices $(\partial f_i/\partial x_j)$ and $(\partial x_i/\partial f_j)$ are inverse, and in particular the Jacobian $\mathrm{j} = \det(\partial f_j/\partial x_i)$ is non-zero. □

Chapter 6

Polynomial rings and regular local rings

6.1 Regular local rings

Suppose that A is a commutative Noetherian local ring with maximal ideal $\mathfrak{M}$, so that $K = A/\mathfrak{M}$ is a field. Recall from Section 2.3 that the Krull dimension of A is equal to the length s of any system of parameters $x_1, \dots, x_s$ for A, and is also equal to the Krull dimension of $\mathrm{Gr}(A)$. The latter is a graded ring, generated by degree one elements, and so its Krull dimension is at most equal to $\dim_K(\mathfrak{M}/\mathfrak{M}^2)$, with equality if and only if $\mathrm{Gr}(A)$ is a polynomial ring on the degree one generators. We say that A is **regular** if $\dim(A) = \dim_K(\mathfrak{M}/\mathfrak{M}^2)$, or equivalently if $\mathrm{Gr}(A)$ is a polynomial ring. Thus regular local rings are the local analogues of graded polynomial rings. It easily follows from the fact that $\mathrm{Gr}(A)$ is a polynomial ring, that A is an integral domain.

In the next section, we prove a theorem of Serre, which states that a commutative Noetherian local ring is regular if and only if it has finite homological dimension. We also prove the graded analogue of this, which states that a commutative graded ring $A = \bigoplus_{j=0}^{\infty} A_j$ with $A_0 = K$ and finitely generated over K by elements of positive degree, is a polynomial ring if and only if it has finite homological dimension. This is Serre's converse to Hilbert's syzygy theorem. We shall use it in Section 7.2 as a way of recognizing graded polynomial rings.

If A is a regular local ring with maximal ideal $\mathfrak{M}$, choose elements $x_1, \dots, x_s \in \mathfrak{M}$ whose images form a basis for $\mathfrak{M}/\mathfrak{M}^2$. Now because $\mathrm{Gr}(A)$ is a polynomial ring, x_1 is not a zero divisor in A. Moreover, $A/(x_1)$ is again a regular local ring, because $\dim_K(\mathfrak{M}/(\mathfrak{M}^2, x_1)) = s - 1 = \dim(A/(x_1))$. So by induction, we see that $x_1, \dots, x_s$ is a regular sequence. So a regular local ring is Cohen–Macaulay.

Next, we observe that in the proof of Lemma 4.2.1, all that was used was the fact that $x_1, \dots, x_s$ was a regular sequence, and $A/(x_1, \dots, x_s) = K$. So exactly the same proof shows that in our situation, the Koszul complex $\mathfrak{K}^A(x_1,\dots,x_s)$ is a free resolution of K as an A-module. So a regular local ring of Krull dimension s also has homological dimension s, and

$$\dim_K \operatorname{Tor}_i^A(K,K) = \binom{s}{i}.$$

We summarize what we know so far.

Theorem 6.1.1 *If A is a regular local ring of Krull dimension s, with maximal ideal $\mathfrak{M}$ and $K = A/\mathfrak{M}$, then*
(i) $\mathrm{Gr}(A) \cong K[x_1,\dots,x_s]$,
(ii) *A is an integral domain,*
(iii) *A is Cohen–Macaulay,*
(iv) *A has global dimension s, and*
(v) $\dim_K \operatorname{Tor}_i^A(K,K) = \binom{s}{i}$. □

Finally, we observe that the proofs of Theorem 4.4.4, Theorem 4.5.1 and Corollary 4.5.2 work just as well for regular local rings as for graded polynomial rings.

Theorem 6.1.2 (Auslander–Buchsbaum) *If M is a finitely generated module for a regular local ring A of Krull dimension s, then*

$$\operatorname{hdim}_A(M) + \operatorname{hcodim}_A(M) = s.$$

□

Theorem 6.1.3 (Ischebeck) *Suppose that A is a regular local ring of Krull dimension s, and M and N are finitely generated A-modules. Then $\operatorname{Ext}^i_A(N,M) = 0$ for $i < \operatorname{depth}(M) - \dim(N)$.* □

Corollary 6.1.4 *If M is a Cohen–Macaulay graded module of Krull dimension n for a regular local ring A of Krull dimension s then $\operatorname{Ext}^i_A(M,A) = 0$ for $i \neq s-n$.* □

6.2 Serre's converse to Hilbert's syzygy theorem

Lemma 6.2.1 *Under Hypothesis 4.3.2, if $\dim_K(\mathfrak{M}/\mathfrak{M}^2) = s$ then*

$$\dim_K \operatorname{Tor}_i^A(K,K) \geq \binom{s}{i}.$$

Proof We prove this using the Koszul complex (see Section 4.2). Choose elements $x_1, \dots, x_s \in \mathfrak{M}$ whose images form a basis for $\mathfrak{M}/\mathfrak{M}^2$. Then the complex $\mathfrak{K} = \mathfrak{K}^A(x_1, \dots, x_s)$ is not necessarily exact, because $x_1, \dots, x_s$ is not necessarily a regular sequence, but it does consist of free modules. Moreover, since each x_i lies in $\mathfrak{M}$, the differential on $\mathfrak{K}$ induces a map

$$d : \mathfrak{K}_i/\mathfrak{M}\mathfrak{K}_i \to \mathfrak{M}\mathfrak{K}_{i-1}/\mathfrak{M}^2\mathfrak{K}_{i-1}.$$

Since the x_i are linearly independent modulo $\mathfrak{M}^2$, this map is injective.

Now let

$$\mathbf{F} : \qquad \cdots \to F_1 \to F_0 \to K \to 0$$

be the minimal resolution of K as an A-module. Then the comparison theorem gives us a map of chain complexes

$$\begin{array}{ccccccc} \cdots \to & \mathfrak{K}_1 & \to & \mathfrak{K}_0 & \to & K & \to 0 \\ & \downarrow & & \downarrow & & \| & \\ \cdots \to & F_1 & \to & F_0 & \to & K & \to 0. \end{array}$$

We claim that the induced map

$$\mathfrak{K}_i/\mathfrak{M}\mathfrak{K}_i = H_i(\mathfrak{K}/\mathfrak{M}\mathfrak{K}) \to F_i/\mathfrak{M}F_i = H_i(\mathbf{F}/\mathfrak{M}\mathbf{F}) = \mathrm{Tor}_i^A(K, K)$$

is injective; or equivalently that the free module $\mathfrak{K}_i$ is injected as a summand of the free module F_i. For $i = 0$, it is clearly an isomorphism of one dimensional vector spaces over K. For $i > 0$, we work by induction. Since $\mathfrak{K}_{i-1}$ is a summand of F_{i-1}, we have a diagram

$$\begin{array}{ccc} \mathfrak{K}_i/\mathfrak{M}\mathfrak{K}_i & \to & F_i/\mathfrak{M}F_i \\ \downarrow & & \downarrow \\ \mathfrak{M}\mathfrak{K}_{i-1}/\mathfrak{M}^2\mathfrak{K}_{i-1} & \to & \mathfrak{M}F_{i-1}/\mathfrak{M}^2F_{i-1} \end{array}$$

Since the left hand vertical map and the bottom horizontal map are injective, it follows that the top horizontal map is injective. This completes the proof by induction. Finally, we observe that

$$\dim_K(\mathfrak{K}_i/\mathfrak{M}\mathfrak{K}_i) = \binom{s}{i}$$

so that the lemma is proved. □

Theorem 6.2.2 (Serre)
Assume Hypothesis 4.3.2, and suppose that $\mathrm{Tor}_s^A(K, K) \neq 0$ *while* $\mathrm{Tor}_j^A(K, K) = 0$ *for* $j > s$ *(note that in particular this is true for some* s *if* A *has finite global dimension). Then in case* (a)*,* A *is a regular local ring of Krull dimension* s*, while in case* (b) A *is a polynomial ring on* s *generators of positive degree.*

Proof The proof of Theorem 4.4.4 works unaltered if we replace a polynomial ring by a ring satisfying our hypotheses, and proves that for any finitely generated graded module M we have

$$\mathrm{hdim}_A(M) + \mathrm{hcodim}_A(M) = s.$$

We apply this with $M = A$. Since $\mathrm{hdim}_A(A) = 0$, we see (using Theorem 4.4.3) that A has depth s. So we have

$$s = \mathrm{depth}(A) \leq \text{Krull dimension of } A \leq \dim_K(\mathfrak{M}/\mathfrak{M}^2).$$

On the other hand, the above lemma shows that the global dimension s is at least as big as $\dim_K(\mathfrak{M}/\mathfrak{M}^2)$. It follows that the above inequalities are equalities, so that the Krull dimension of A is equal to $\dim_K(\mathfrak{M}/\mathfrak{M}^2)$. In case (a) this shows that A is regular, while in case (b) we deduce from this that A is a polynomial ring. □

Corollary 6.2.3 *Suppose that $A = K[y_1, \dots, y_s]$ is a graded polynomial ring over K on homogeneous generators y_i of positive degree. If A' is a finitely generated graded K-subalgebra of A, over which A is a free module, then A' is a polynomial ring.*

Proof By the theorem, it suffices to show that A' has finite global dimension. Since A is a free A'-module, tensoring over A' with A is exact. So if M is an A'-module, with free resolution

$$\cdots \to F_2 \to F_1 \to F_0 \to M \to 0$$

then tensoring over A' with A, we obtain a free resolution

$$\cdots \to A \otimes_{A'} F_2 \to A \otimes_{A'} F_1 \to A \otimes_{A'} F_0 \to A \otimes_{A'} M \to 0$$

of $A \otimes_{A'} M$ as an A-module. So we have

$$\mathrm{Tor}_i^{A'}(M, K) \cong \mathrm{Tor}_i^A(A \otimes_{A'} M, K).$$

In particular, since the Hilbert syzygy theorem says the right hand side vanishes for $i > s$, so does the left hand side, and so M has homological dimension at most s. Therefore A' has global dimension at most s, and so by the theorem it is a polynomial ring. □

Corollary 6.2.4 (Serre) *Any localization of a regular local ring is again a regular local ring.*

Proof Suppose that A is a commutative Noetherian local ring. Any finitely generated A-module has a projective resolution of length at most s, if and only if $\mathrm{Ext}_A^i(M, N)$ vanishes for all $i > s$ and all finitely generated A-modules M and N. By Lemma 2.2.2 and repeated use of the long exact sequence in Ext, this happens if and only if $\mathrm{Ext}^i_A(A/\mathfrak{p}, N)$ vanishes for all $i > s$, all prime ideals $\mathfrak{p}$, and all A-modules N. This happens if and only if $A/\mathfrak{p}$ has a projective resolution of length at most s for all prime ideals $\mathfrak{p}$.

Now recall from Section 1.4 that localization is exact, and use the fact that a local ring A is regular if and only if every finitely generated A-module has a projective resolution of finite length (Theorems 6.1.1 and 6.2.2). □

6.3 Uniqueness of factorization

In this section, we prove a theorem of Auslander and Buchsbaum [9], which states that a regular local ring is a unique factorization domain. We begin with a lemma.

Lemma 6.3.1 *Suppose that A is a Noetherian integral domain. If $x \in A$ is prime and $A[x^{-1}]$ is a unique factorization domain, then A is also a unique factorization domain.*

Proof If $a \in A$ is not divisible by x, and a factorizes in $A[x^{-1}]$, then for some $r > 0$, ax^r factorizes in A in such a way that neither factor is a power of x. So if $a \neq x$ is irreducible in A then it is irreducible in $A[x^{-1}]$ and hence prime in $A[x^{-1}]$. So if a divides bc in A, then a divides b or c, say b, in $A[x^{-1}]$, and hence for some $r \geq 0$, a divides bx^r in A. If $r > 0$, then x divides $(bx^r/a).a$ and does not divide a, so x divides bx^r/a, and hence a divides bx^{r-1}. Continuing this way, we see that a divides b, and so a is prime. □

Proposition 6.3.2 (Auslander and Buchsbaum) *If A is a regular local ring then A is a unique factorization domain.*

Proof We give a version (Matsumura [59] Theorem 20.3) of a proof by Kaplansky. We prove the theorem by induction on $\dim(A)$. Let $\mathfrak{M}$ be the maximal ideal of A. If $\dim(A) = 0$ then $\mathfrak{M} = 0$ and A is a field, so we may assume that $\dim(A) > 0$. Choose $x \in \mathfrak{M}$ with $x \notin \mathfrak{M}^2$. Since $A/(x)$ is an integral domain (indeed, it is a regular local ring), x is prime in A. So it suffices, by the lemma, to prove that $A[x^{-1}]$ is a unique factorization domain. To do this, it suffices to prove that any height one prime ideal is principal: for then if a is irreducible, a minimal prime ideal $\mathfrak{P}$ containing a has height one by Corollary 2.3.3, and is hence principal, say $\mathfrak{P} = (b)$; then $a = bc$ for some c, and since a is irreducible, c is invertible and $\mathfrak{P} = (a)$ so that a is prime.

We now suppose that $\mathfrak{P}$ is a height one prime ideal in $A[x^{-1}]$, and prove that $\mathfrak{P}$ is principal. Set $\mathfrak{p} = \mathfrak{P} \cap A$, so that $\mathfrak{P} = \mathfrak{p}A[x^{-1}]$. Since A is regular, $\mathfrak{p}$ has a finite free resolution as an A-module, so that since localization is exact (Section 1.4), $\mathfrak{P}$ has a finite free resolution as an $A[x^{-1}]$-module.

Now $A[x^{-1}]$ is no longer a local ring (because the prime ideals correspond to the prime ideals in A not containing x, so that $\mathfrak{M}$ is not included), but if $\mathfrak{Q}$ is any prime ideal in $A[x^{-1}]$ then the localization $A[x^{-1}]_{\mathfrak{Q}} = A_{A\cap\mathfrak{Q}}$ is a regular local ring (by Corollary 6.2.4) of smaller Krull dimension than A, and is hence a unique factorization domain by the inductive hypothesis. So $\mathfrak{P}_{\mathfrak{Q}}$ is a free $A[x^{-1}]_{\mathfrak{Q}}$-module. So if $M \to M'$ is a surjective map of finitely generated $A[x^{-1}]$-modules, then the cokernel of $\mathrm{Hom}_A(\mathfrak{P}, M) \to \mathrm{Hom}_A(\mathfrak{P}, M')$ has zero localization at any prime $\mathfrak{Q}$, hence has no associated primes, and is hence the zero module, so this map is surjective. Applying this with $M' = \mathfrak{P}$, we see that $\mathfrak{P}$ is a projective $A[x^{-1}]$-module. Since it also has a finite free resolution, we deduce that for some $m \geq 0$, we have an isomorphism $\phi : A[x^{-1}]^m \to \mathfrak{P} \oplus A[x^{-1}]^{m-1}$. Regard $\mathfrak{P}$ as a submodule of $A[x^{-1}]$, so that ϕ may be regarded as a map on $A[x^{-1}]^m$ with image $\mathfrak{P} \oplus A[x^{-1}]^{m-1}$. The mth exterior power of this map is the map $(\det\phi).1$ on $\Lambda^m(A[x^{-1}]^m) = A[x^{-1}]$, and has image $\mathfrak{P}$, so $\mathfrak{P}$ is the principal ideal generated by $\det\phi$. □

6.4 Reflexive modules

Lemma 6.4.1 *Under Hypothesis 4.3.2, suppose that A is an integral domain, $\mathfrak{M} \neq 0$, and $M \neq 0$ is a reflexive A-module. Then M has depth at least one. If, furthermore, A has depth at least two, then so does M.*

Proof If x is a non-zero element of $\mathfrak{M}$, $x \notin \mathfrak{M}^2$, then x is not a zero divisor on M, since M is torsion-free (see Section 3.4). So M has depth at least one.

Now suppose that A has depth at least two. We have $\mathrm{Hom}_A(K, M) = 0$, so by Theorem 4.4.3, M has depth at least two if and only if $\mathrm{Ext}^1_A(K, M) = 0$. If $0 \to M \to E \to K \to 0$ is a non-split short exact sequence, then E is torsion-free, and so $E \to E^{**}$ is injective. Applying $\mathrm{Hom}_A(-, A)$ to this sequence, we get

$$0 \to E^* \to M^* \to \mathrm{Ext}^1_A(K, A) = 0$$

since A has depth at least two. So $M \to M^{**} \to E^{**}$ are isomorphisms, which is absurd as $E \to E^{**}$ is injective. □

Lemma 6.4.2 *Suppose that M is a finitely generated module over a local integral domain A with maximal ideal $\mathfrak{M}$ and $K = A/\mathfrak{M}$. Then M is free if and only if the natural map*

$$\psi : M \otimes_A M^* \to \mathrm{Hom}_A(M, M)$$

given by $\psi(m \otimes f)(m') = f(m')m$ is an isomorphism.

Proof If M is free, it is easy to see using matrices over A that ψ is an isomorphism. Conversely, suppose that ψ is an isomorphism. Choose an element $\sum_i m_i \otimes f_i \in M \otimes M^*$ whose image under ψ is the identity homomorphism. Choose i with $\psi(m_i \otimes f_i) \notin \mathfrak{M}.\mathrm{Hom}_A(M, M)$. Then for some $m' \in M$,

$$f_i(m')m_i = \psi(m_i \otimes f_i)(m') \notin \mathfrak{M}.M$$

and so $f_i(m') \notin \mathfrak{M}$. So the map $f_i : M \to A$ is surjective. Since A is a free A-module, this implies that $M \cong A \oplus M_1$ for some finitely generated A-module M_1. It is easy to see that the restriction of ψ to $M_1 \otimes_A M_1^* \to \mathrm{Hom}_A(M_1, M_1)$ must again be an isomorphism. Since M is finitely generated, we may continue this way by induction until we reach the conclusion that M is free. □

Theorem 6.4.3 (Auslander [8]) *Suppose that M is a finitely generated reflexive module over a regular local ring A with maximal ideal $\mathfrak{M}$, and $K = A/\mathfrak{M}$. If $\mathrm{Hom}_A(M, M)$ is isomorphic to a direct sum of copies of M, then M is free.*

Proof We prove this theorem by induction on $\dim(A)$. If $\dim(A) \leq 2$ then every reflexive A-module is free by Lemma 6.4.1, so assume that $\dim(A) \geq 3$. If $\mathfrak{p}$ is a prime ideal in A then by Corollary 6.2.4 $A_\mathfrak{p}$ is a regular local ring, $M_\mathfrak{p}$ is a reflexive $A_\mathfrak{p}$-module, and $\mathrm{Hom}_{A_\mathfrak{p}}(M_\mathfrak{p}, M_\mathfrak{p})$ is isomorphic to a direct sum of copies of $M_\mathfrak{p}$. So we may assume that $M_\mathfrak{p}$ is $A_\mathfrak{p}$-free for every non-maximal prime ideal $\mathfrak{p}$ by induction.

We begin with the case $\dim(A) = 3$. In this case, we shall prove that the natural map

$$\psi : M \otimes M^* \to \mathrm{Hom}_A(M, M)$$

described in Lemma 6.4.2 is an isomorphism, so that by that lemma, M is free. We have exact sequences

$$0 \to \ker \psi \to M \otimes M^* \to \mathrm{Im}\, \psi \to 0$$

$$0 \to \mathrm{Im}\, \psi \to \mathrm{Hom}_A(M, M) \to \mathrm{Coker}\, \psi \to 0.$$

Since ψ induces an isomorphism after localizing at any non-maximal prime ideal, $\ker \psi$ and $\mathrm{Coker}\, \psi$ have only the maximal ideal as associated primes, and hence have Krull dimension zero (i.e., finite length). So by Corollary 6.1.4 we have $\mathrm{Ext}^i_A(\ker \psi, A) = 0$ and $\mathrm{Ext}^i_A(\mathrm{Coker}\, \psi, A) = 0$ for $1 \leq i \leq 2$. It follows that ψ induces isomorphisms

$$\mathrm{Ext}^1_A(M \otimes M^*, A) \cong \mathrm{Ext}^1_A(\mathrm{Im}\, \psi, A) \cong \mathrm{Ext}^1_A(\mathrm{Hom}_A(M, M), A).$$

Also, the localization $\mathrm{Ext}^1_A(\mathrm{Hom}_A(M, M), A)_\mathfrak{p} = 0$ for any non-maximal prime ideal $\mathfrak{p}$, and so $\mathrm{Ext}^1_A(\mathrm{Hom}_A(M, M), A)$ has finite length, and

$$\ell(\mathrm{Ext}^1_A(M \otimes M^*, A)) = \ell(\mathrm{Ext}^1_A(\mathrm{Hom}_A(M, M), A)).$$

By Lemma 6.4.1, M has depth at least two, so by Theorem 6.1.2, M has homological dimension at most one. In other words, M has a free resolution of the form $0 \to F_1 \to F_0 \to M \to 0$. Taking homomorphisms into A, we obtain

$$0 \to M^* \to F_0^* \to F_1^* \to \mathrm{Ext}^1_A(M, A) \to 0.$$

Let Y be the cokernel of $M^* \to F_0^*$, so that we have an exact sequence

$$0 \to M^* \to F_0^* \to Y \to 0.$$

Then we have $\mathrm{Tor}_1^A(M, M^*) \cong \mathrm{Tor}_2^A(M, Y) = 0$, since M has homological dimension at most one, and so we have an exact sequence

$$0 \to F_1 \otimes M^* \to F_0 \otimes M^* \to M \otimes M^* \to 0.$$

This gives us a diagram

$$\begin{array}{ccccc}
\mathrm{Hom}_A(F_0 \otimes M^*, A) & \to & \mathrm{Hom}_A(F_1 \otimes M^*, A) & \to & \mathrm{Ext}_A^1(M \otimes M^*, A) \\
\| & & \| & & \\
\mathrm{Hom}_A(F_0, M^{**}) & \to & \mathrm{Hom}_A(F_1, M^{**}) & & \\
\| & & \| & & \\
\mathrm{Hom}_A(F_0, M) & \to & \mathrm{Hom}_A(F_1, M) & \to & \mathrm{Ext}_A^1(M, M) \to 0 \\
\| & & \| & & \\
F_0^* \otimes M & \to & F_1^* \otimes M & \to & \mathrm{Ext}_A^1(M, A) \otimes M \to 0
\end{array}$$

and so we have an injective map

$$\mathrm{Ext}_A^1(M, A) \otimes M \cong \mathrm{Ext}_A^1(M, M) \to \mathrm{Ext}_A^1(M \otimes M^*, A).$$

So we have

$$\begin{aligned}
\ell(\mathrm{Ext}_A^1(M, A) \otimes M) &\leq \ell(\mathrm{Ext}_A^1(M \otimes M^*, A)) = \ell(\mathrm{Ext}_A^1(\mathrm{Hom}_A(M, M), A)) \\
&= \mathrm{rank}(M).\ell(\mathrm{Ext}_A^1(M, A)) \\
&= (\mathrm{rank}(F_0) - \mathrm{rank}(F_1)).\ell(\mathrm{Ext}_A^1(M, A)).
\end{aligned}$$

Here, we have used $\mathrm{rank}(-)$ to denote $\dim_L(L \otimes_A -)$, where L is the field of fractions of A. Since M is reflexive we have $\mathrm{rank}(M) \neq 0$, so that $\mathrm{Hom}_A(M, M)$ is isomorphic to a direct sum of $\mathrm{rank}(M)$ copies of M.

Now examining the sequence

$$0 \to \mathrm{Tor}_1^A(\mathrm{Ext}_A^1(M, A), M) \to \mathrm{Ext}_A^1(M, A) \otimes F_1 \to$$

$$\mathrm{Ext}_A^1(M, A) \otimes F_0 \to \mathrm{Ext}_A^1(M, A) \otimes M \to 0$$

we see that this implies that $\mathrm{Tor}_1^A(\mathrm{Ext}_A^1(M, A), M) = 0$. On the other hand, since $\mathrm{hdim}_A(M) \leq 1$, if M is not free then $\mathrm{Ext}_A^1(M, A)$ is a non-zero module of finite length, and hence contains K as a submodule. Since F_1 is not a direct summand of F_0, $K \otimes F_1 \to K \otimes F_0$ is not injective, so $\mathrm{Tor}_1^A(K, M) \neq 0$ and

$$\mathrm{Tor}_1^A(K, M) \to \mathrm{Tor}_1^A(\mathrm{Ext}_A^1(M, A), M)$$

is injective. So M must be free, and the case $\dim(A) = 3$ is complete.

Now assume that $\dim(A) \geq 4$. Choose $x \in \mathfrak{M}$, $x \notin \mathfrak{M}^2$, and set $\bar{A} = A/(x)$, $\bar{M} = M/xM$. Since M is reflexive, x is not a zero divisor on M, so we have a short exact sequence

$$0 \to M \xrightarrow{x} M \to \bar{M} \to 0$$

and hence an exact sequence

$$0 \to \mathrm{Hom}_A(M, M) \xrightarrow{x} \mathrm{Hom}_A(M, M) \to \mathrm{Hom}_A(M, \bar{M}) \to \mathrm{Ext}_A^1(M, M).$$

Since $\mathrm{Hom}_A(M, M)$ is isomorphic to a direct sum of $\mathrm{rank}(M)$ copies of M, the cokernel of multiplication by x is a direct sum of this number of copies of $\bar{M}$. Also, $M_\mathfrak{p}$ is free for all non-maximal prime ideals $\mathfrak{p}$, so $\mathrm{Ext}_A^1(M, M)$ has finite length. Furthermore, $\mathrm{Hom}_A(M, \bar{M}) \cong \mathrm{Hom}_{\bar{A}}(\bar{M}, \bar{M})$ and so the above exact sequence gives us

$$0 \to \mathrm{rank}(M).\bar{M} \xrightarrow{\lambda} \mathrm{Hom}_{\bar{A}}(\bar{M}, \bar{M}) \to T \to 0$$

where T has finite length, and in particular $\mathrm{Hom}_{\bar{A}}(T, \bar{A}) = 0$. Since $\dim(\bar{A}) \geq 2$ (in fact $\dim(\bar{A}) \geq 3$), Theorem 6.1.3 implies that $\mathrm{Ext}_{\bar{A}}^1(T, \bar{A}) = 0$, and so taking duals as $\bar{A}$-modules, we see that

$$\lambda^* : \mathrm{Hom}_{\bar{A}}(\bar{M}, \bar{M})^* \to \mathrm{rank}(M).\bar{M}^*$$

is an isomorphism. Dualizing again and using Lemma 3.4.1 (v), we see that

$$\mathrm{rank}(M).\bar{M}^{**} \cong \mathrm{Hom}_{\bar{A}}(\bar{M}^{**}, \bar{M}^{**}).$$

By Lemma 3.4.1 (iv), $\bar{M}^{**}$ is reflexive, so by the inductive hypothesis, $\bar{M}^{**}$ is a free $\bar{A}$-module. So $\bar{M}^* \cong \bar{M}^{***}$ is also a free $\bar{A}$-module.

Now M has depth at least two by Lemma 6.4.1, so $\bar{M}$ is torsion-free and hence $\bar{M} \to \bar{M}^{**}$ is injective by Lemma 3.4.1 (iii). Moreover, $\bar{M} \to \bar{M}^{**}$ induces an isomorphism after localizing at any non-maximal prime ideal $\mathfrak{p}$, so its cokernel X has finite length. Since $\dim(\bar{A}) \geq 3$, Theorem 6.1.3 implies that $\mathrm{Ext}_{\bar{A}}^2(X, \bar{A}) = 0$. Since $\bar{M}^{**}$ is free, $\mathrm{Ext}_{\bar{A}}^1(\bar{M}^{**}, \bar{A}) = 0$, so the exact sequence

$$\cdots \to \mathrm{Ext}_{\bar{A}}^1(\bar{M}^{**}, \bar{A}) \to \mathrm{Ext}_{\bar{A}}^1(\bar{M}, \bar{A}) \to \mathrm{Ext}_{\bar{A}}^2(X, \bar{A}) \to \cdots$$

shows that $\mathrm{Ext}_{\bar{A}}^1(\bar{M}, \bar{A}) = 0$.

Now look at the exact sequence

$$0 \to M^* \xrightarrow{x} M^* \to \bar{M}^* \to \mathrm{Ext}_A^1(M, A) \xrightarrow{x} \mathrm{Ext}_A^1(M, A) \to \mathrm{Ext}_A^1(M, \bar{A}).$$

The last term in this sequence is equal to $\mathrm{Ext}^1_{\bar{A}}(\bar{M}, \bar{A})$, which we have just shown to be zero, so multiplication by x is surjective on $\mathrm{Ext}^1_A(M, A)$. Since $x \in \mathfrak{M}$, this implies that $\mathrm{Ext}^1_A(M, A) = 0$, so that we have an exact sequence

$$0 \to M^* \xrightarrow{x} M^* \to \bar{M}^* \to 0.$$

Since $\bar{M}^*$ is $\bar{A}$-free, it has a resolution of length one as an A-module, so the long exact Ext sequence shows that M^* is A-free. Hence $M \cong M^{**}$ is A-free. □

Chapter 7

Groups Generated by Pseudoreflections

7.1 Reflections and pseudoreflections

Suppose that V is a finite dimensional vector space over a field K. Recall from Section 2.6 that a **pseudoreflection** is a linear automorphism of V of finite order whose fixed points have codimension one. If the automorphism is diagonalizable, this is the same as saying that all but one of the eigenvalues are equal to one. A **reflection** is a diagonalizable pseudoreflection of order two. In this case, the remaining eigenvalue is equal to -1.

If G_1 and G_2 are generated by pseudoreflections on vector spaces V_1 and V_2 respectively, then $G_1 \times G_2$ is generated by pseudoreflections on the space $V_1 \oplus V_2$. If G is generated by pseudoreflections on V and cannot be decomposed in this way, then G is said to be an indecomposable pseudoreflection group.

If $K = \mathbb{Q}$, the field of rational numbers, then all pseudoreflections are reflections. The product of two distinct reflections is a planar rotation, and since its trace is rational, this product has order two, three, four or six. The finite groups generated by reflections in this case are called the **crystallographic reflection groups**. The indecomposable ones are in one–one correspondence with the Dynkin diagrams A_n, B_n, D_n, E_6, E_7, E_8, F_4 and G_2 (the reflection groups corresponding to C_n are the same as those for B_n; only the root lengths are different). For a further discussion of the crystallographic reflection groups, see Humphreys [50], Chapter III, and Bourbaki [17].

Over $K = \mathbb{R}$, again all pseudoreflections are reflections, and the classification of finite groups generated by reflections was obtained in this case by Coxeter [28] (see also Grove and Benson [40]). As well as the above groups, one also has the dihedral groups acting on $\mathbb{R}^2$ in the usual way (A_2, B_2 and G_2 in fact correspond to the dihedral groups of orders six, eight and twelve), the direct product of the alternating group A_5 with $\mathbb{Z}/2$, acting on $\mathbb{R}^3$ as the symmetries of an icosahedron, and a group containing the double cover of $A_5 \times A_5$ as a subgroup of index two, acting on $\mathbb{R}^4$. The latter may be described as follows. The inverse image of A_5 under the double cover map $SU(2) \to SO(3)$ is a proper double cover of A_5, written $2A_5$, and called the binary icosahedral group.

$$\begin{array}{ccccccccc} 1 \to & \mathbb{Z}/2 & \to & SU(2) & \to & SO(3) & \to 1 \\ & \| & & \cup & & \cup & \\ 1 \to & \mathbb{Z}/2 & \to & 2A_5 & \to & A_5 & \to 1 \end{array}$$

Now there is also a double cover map $SU(2) \times SU(2) \to SO(4)$, and the image of $2A_5 \times 2A_5$ under this map is the quotient by the diagonal of the two central elements, and is hence a diagonal double cover of $A_5 \times A_5$.

$$\begin{array}{ccccccccc} 1 \to & \mathbb{Z}/2 & \to & SU(2) \times SU(2) & \to & SO(4) & \to 1 \\ & \| & & \cup & & \cup & \\ 1 \to & \mathbb{Z}/2 & \to & 2A_5 \times 2A_5 & \to & 2(A_5 \times A_5) & \to 1 \end{array}$$

The group $O(4)$ is a split extension of $SO(4)$ by an element of order two swapping the two copies of $SU(2)$, and adjoining this automorphism to $2(A_5 \times A_5)$ gives the desired reflection group.

Over the complexes $K = \mathbb{C}$, the situation is much more complicated. Nevertheless, Shephard and Todd [95] succeeded in classifying the groups generated by pseudoreflections, and Clark and Ewing [27] used this to classify the groups generated by pseudoreflections over an arbitrary field of characteristic coprime to the group order. A list of the indecomposable pseudoreflection groups is given in the table on page 81, which is taken from Clark and Ewing [27]. The numbering of the groups corresponds to the numbering used by Shephard and Todd [95].

A few words of explanation concerning this table are in order. We shall see in the next section that for a group generated by pseudoreflections, the invariants form a polynomial ring. The fourth column lists the degrees of the polynomial generators. Clark and Ewing double the degrees in their list, because they are interested in topological applications where the vector space V occurs in degree two, so the degrees in our table are half of theirs.

The next column describes the field generated by the character values of the representation; here, ζ_n denotes $e^{2\pi i/n}$, a primitive nth root of unity. Now in general, a complex representation of a finite group is not necessarily equivalent to a representation whose matrix entries are in the field generated by the character values, and the theory of the Schur index measures the extent to which this fails. However, for a group generated by pseudoreflections, the Schur index is always one, as the following proposition shows.

	dim	$\|G\|$	degrees	field	primes
1	n	$(n+1)!$	$2, 3, \ldots, n+1$	$\mathbf{Q}$	$> n+1$
2a	n	$qm^{n-1}n!$ $(q\|m$ and $m > 1)$	$m, 2m, \ldots, (n-1)m, qn$	$\mathbf{Q}(\zeta_m)$	$1 \pmod{m}, > n$
2b	2	$2m\ (m > 2)$	$2, m$	$\mathbf{Q}(\zeta_m + \zeta_m^{-1})$	$\pm 1 \pmod{m}$
3	1	m	m	$\mathbf{Q}(\zeta_m)$	$1 \pmod{m}$
4	2	$24 = 2^3.3$	$4, 6$	$\mathbf{Q}(\zeta_3)$	$1 \pmod{3}$
5	2	$72 = 2^3.3^2$	$6, 12$	$\mathbf{Q}(\zeta_3)$	$1 \pmod{3}$
6	2	$48 = 2^4.3$	$4, 12$	$\mathbf{Q}(\zeta_{12})$	$1 \pmod{12}$
7	2	$144 = 2^4.3^2$	$12, 12$	$\mathbf{Q}(\zeta_{12})$	$1 \pmod{12}$
8	2	$96 = 2^5.3$	$8, 12$	$\mathbf{Q}(i)$	$1 \pmod{4}$
9	2	$192 = 2^6.3$	$8, 24$	$\mathbf{Q}(\zeta_8)$	$1 \pmod{8}$
10	2	$288 = 2^5.3^2$	$12, 24$	$\mathbf{Q}(\zeta_{12})$	$1 \pmod{12}$
11	2	$576 = 2^6.3^2$	$24, 24$	$\mathbf{Q}(\zeta_{24})$	$1 \pmod{24}$
12	2	$48 = 2^4.3$	$6, 8$	$\mathbf{Q}(\sqrt{-2})$	$1, 3 \pmod{8}, \neq 3$
13	2	$96 = 2^5.3$	$8, 12$	$\mathbf{Q}(\zeta_8)$	$1 \pmod{8}$
14	2	$144 = 2^4.3^2$	$6, 24$	$\mathbf{Q}(\zeta_3, \sqrt{-2})$	$1, 19 \pmod{24}$
15	2	$288 = 2^5.3^2$	$12, 24$	$\mathbf{Q}(\zeta_{24})$	$1 \pmod{24}$
16	2	$600 = 2^3.3.5^2$	$20, 30$	$\mathbf{Q}(\zeta_5)$	$1 \pmod{5}$
17	2	$1200 = 2^4.3.5^2$	$20, 60$	$\mathbf{Q}(\zeta_{20})$	$1 \pmod{20}$
18	2	$1800 = 2^3.3^2.5^2$	$30, 60$	$\mathbf{Q}(\zeta_{15})$	$1 \pmod{15}$
19	2	$3600 = 2^4.3^2.5^2$	$60, 60$	$\mathbf{Q}(\zeta_{60})$	$1 \pmod{60}$
20	2	$360 = 2^3.3^2.5$	$12, 30$	$\mathbf{Q}(\zeta_3, \sqrt{5})$	$1, 4 \pmod{15}$
21	2	$720 = 2^4.3^2.5$	$12, 60$	$\mathbf{Q}(\zeta_{12}, \sqrt{5})$	$1, 49 \pmod{60}$
22	2	$240 = 2^4.3.5$	$12, 20$	$\mathbf{Q}(i, \sqrt{5})$	$1, 9 \pmod{20}$
23	3	$120 = 2^3.3.5$	$2, 6, 10$	$\mathbf{Q}(\sqrt{5})$	$1, 4 \pmod{5}$
24	3	$336 = 2^4.3.7$	$4, 6, 14$	$\mathbf{Q}(\sqrt{-7})$	$1, 2, 4 \pmod{7}$
25	3	$648 = 2^3.3^4$	$6, 9, 12$	$\mathbf{Q}(\zeta_3)$	$1 \pmod{3}$
26	3	$1296 = 2^4.3^4$	$6, 12, 18$	$\mathbf{Q}(\zeta_3)$	$1 \pmod{3}$
27	3	$2160 = 2^4.3^3.5$	$6, 12, 30$	$\mathbf{Q}(\zeta_3, \sqrt{5})$	$1, 4 \pmod{15}$
28	4	$1152 = 2^7.3^2$	$2, 6, 8, 12$	$\mathbf{Q}$	$\neq 2$ or 3
29	4	$7680 = 2^9.3.5$	$4, 8, 12, 20$	$\mathbf{Q}(i)$	$1 \pmod{4}, \neq 5$
30	4	$2^6.3^2.5^2$	$2, 12, 20, 30$	$\mathbf{Q}(\sqrt{5})$	$1, 4 \pmod{5}$
31	4	$2^{10}.3^2.5$	$8, 12, 20, 24$	$\mathbf{Q}(i)$	$1 \pmod{4}, \neq 5$
32	4	$2^7.3^5.5$	$12, 18, 24, 30$	$\mathbf{Q}(\zeta_3)$	$1 \pmod{3}$
33	5	$2^7.3^4.5$	$4, 6, 10, 12, 18$	$\mathbf{Q}(\zeta_3)$	$1 \pmod{3}$
34	6	$2^9.3^7.5.7$	$6, 12, 18, 24, 30, 42$	$\mathbf{Q}(\zeta_3)$	$1 \pmod{3}, \neq 7$
35	6	$2^7.3^4.5$	$2, 5, 6, 8, 9, 12$	$\mathbf{Q}$	$\neq 2, 3$ or 5
36	7	$2^{10}.3^4.5.7$	$2, 6, 8, 10, 12, 14, 18$	$\mathbf{Q}$	$\neq 2, 3, 5$ or 7
37	8	$2^{14}.3^5.5^2.7$	$2, 8, 12, 14, 18, 20, 24, 30$	$\mathbf{Q}$	$\neq 2, 3, 5$ or 7

Table 7.1: Finite groups generated by pseudoreflections

Proposition 7.1.1 *If G acts on a complex vector space V as a finite group generated by pseudoreflections, then V is equivalent to a representation whose matrix entries lie in the subfield of $\mathbf{C}$ generated by the character values of G on V.*

Proof Let K be the subfield of $\mathbf{C}$ generated by the character values of G on V. Then there is a Wedderburn component of the group algebra KG corresponding to V, of the form $\Lambda \cong \mathrm{Mat}_m(\Delta)$, where Δ is a division ring whose center is K. Let K' be a maximal subfield of Δ, so that $|\Delta : K'| = |K' : K|$, and regard K' as embedded in Λ as diagonal matrices. If V' is the (unique) irreducible Λ-module, then $\mathbf{C} \otimes_{K'} V' \cong V$. Now the fixed points of an element $g \in G$ on V' form a Δ-linear subspace, and hence the codimension over K' is divisible by $|\Delta : K'|$. In particular, if g is a pseudoreflection then this codimension is one, and so $\Delta = K' = K$, $\Lambda \cong \mathrm{Mat}_n(K)$, and $\mathbf{C} \otimes_K V' \cong V$. □

The next column of the table gives the values of p for which $|G|$ is not divisible by p and the field of character values lies in $\mathbf{Q}_p^\wedge$, the p-adic numbers, so that V is equivalent to a representation with entries in $\mathbf{Q}_p^\wedge$. The relevance of this depends on the following lemma.

Lemma 7.1.2 *If G is a finite group of order coprime to p generated by pseudoreflections on an n-dimensional vector space $\bar{V}$ over $\mathbf{F}_p$, then there is an action of G on a free $\mathbf{Z}_p^\wedge$-module $\hat{V}$ of rank n with the following properties:*

(i) $\bar{V} \cong \mathbf{F}_p \otimes_{\mathbf{Z}_p^\wedge} \hat{V}$ as a representation of G, and

(ii) $V = \mathbf{Q}_p^\wedge \otimes_{\mathbf{Z}_p^\wedge} \hat{V}$ is a $\mathbf{Q}_p^\wedge$-vector space of dimension n on which G is generated by pseudoreflections.

Furthermore, the $\mathbf{Z}_p^\wedge G$-module $\hat{V}$ is determined up to isomorphism by the $\mathbf{F}_pG$-module $\bar{V}$.

Proof We lift inductively to an action on a free $\mathbf{Z}/p^i$-module of rank n as follows. The kernel of $GL_n(\mathbf{Z}/p^i) \to GL_n(\mathbf{Z}/p^{i-1})$ is an elementary abelian p-group $(\mathbf{Z}/p)^{n^2}$, which as a representation of G is equivalent to $\bar{V}^* \otimes \bar{V}$. Since $|G|$ is coprime to p, $H^2(G, \bar{V}^* \otimes \bar{V}) = 0$, so there is no obstruction to lifting a map $G \to GL_n(\mathbf{Z}/p^{i-1})$ to a map $G \to GL_n(\mathbf{Z}/p^i)$. Similarly, $H^1(G, \bar{V}^* \otimes \bar{V}) = 0$, so such a lift is unique up to conjugacy. Since $GL_n(\mathbf{Z}_p^\wedge) = \varprojlim_i GL_n(\mathbf{Z}/p^i)$, we may lift to a map $G \to GL_n(\mathbf{Z}_p^\wedge)$, unique up to conjugacy. Applying this in particular to cyclic subgroups of G, we see that dimensions of fixed spaces are preserved, and so G is generated by pseudoreflections on V. □

7.2 The Shephard–Todd theorem

The main theorem of this section states that in coprime characteristic, the invariants form a polynomial ring if and only if the group is generated by pseudoreflections.

This theorem was originally proved over the complex numbers by Shephard and Todd [95], using their classification of the finite groups generated by complex pseudoreflections. Later, Chevalley [26] gave a proof in the real case without using the classification. Serre pointed out how to modify Chevalley's proof so that it works in the complex case. Following Smith [96] and Bourbaki [17] Chapter 5 §5, Exercise 7, we shall present a more homological proof which works in general in coprime characteristic.

Theorem 7.2.1 *Suppose that V is a finite dimensional faithful representation of a finite group G over a field K of characteristic coprime to $|G|$. Then we have the implications* (i) $\Leftarrow$ (ii) $\Leftrightarrow$ (iii) $\Leftrightarrow$ (iv) $\Leftrightarrow$ (v) *among the following statements:*

(i) *G is generated by elements which act on V as pseudoreflections.*

(ii) $K[V]^G = K[f_1, \dots, f_n]$ *is a polynomial ring.*

(iii) $K[V]$ *is a free $K[V]^G$-module.*

(iv) $\mathrm{Tor}_1^{K[V]^G}(K, K[V]) = 0$.

(v) *The map*

$$K[V]^G_+ \otimes_{K[V]^G} K[V] \xrightarrow{\phi} K[V],$$

given by multiplication, is injective. Here, $K[V]^G_+$ denotes the ideal in $K[V]^G$ generated by the elements of positive degree.

If the characteristic of K is coprime to $|G|$ then all five statements are equivalent, and if they hold then the number of pseudoreflections in G is equal to $\sum_i (k_i - 1)$, where $k_i = \deg(f_i)$.

Proof (ii) $\Leftrightarrow$ (iii): This follows from Corollary 6.2.3.

(iii) $\Leftrightarrow$ (iv): This follows from the discussion in Section 4.2.

(iv) $\Leftrightarrow$ (v): The exact sequence

$$0 \to K[V]^G_+ \to K[V]^G \to K \to 0$$

gives an exact sequence

$$0 \to \mathrm{Tor}_1^{K[V]^G}(K, K[V]) \to K[V]^G_+ \otimes_{K[V]^G} K[V]$$

$$\to K[V]^G \otimes_{K[V]^G} K[V] \to K \otimes_{K[V]^G} K[V] \to 0.$$

After identifying $K[V]^G \otimes_{K[V]^G} K[V]$ with $K[V]$, the middle map in this sequence is identified with ϕ.

Next, we show that if the characteristic of K is coprime to $|G|$ then (i) $\Rightarrow$ (v). We begin by observing that if $f \in K[V]$ and g is a pseudoreflection, then $g(f)$ agrees with f on the hyperplane V^g. Let $\alpha \in K[V]_1$ be a linear function vanishing on V^g. Since $g(f) - f$ vanishes on V^g we have

$$g(f) - f = \alpha . f'$$

with $f' \in K[V]$ an element of degree one less than that of f. Now if

$$\zeta \in K[V]^G_+ \otimes_{K[V]^G} K[V],$$

then applying the above, we see that

$$g(\zeta) - \zeta = (1 \otimes \alpha).\zeta',$$

with ζ' an element of degree one less than that of ζ.

If ϕ is not mono, choose ζ of minimal degree in $\ker(\phi)$. Since ϕ is a G-map, $g(\zeta)$ is also in $\ker(\phi)$, so

$$0 = \phi((1 \otimes \alpha).\zeta') = \alpha.\phi(\zeta').$$

Now $K[V]$ has no zero divisors, so this implies that $\phi(\zeta') = 0$. Since ζ' has smaller degree than ζ, this implies that $\zeta' = 0$, and so $g(\zeta) = \zeta$. This holds for each pseudoreflection $g \in G$, and since G is generated by pseudoreflections, this implies that ζ is G-invariant. Thus it lies in $K[V]^G_+ \otimes_{K[V]^G} K[V]^G$. But since the characteristic of K is coprime to the order of G, $K[V]^G$ is a direct summand of $K[V]$ as a $K[V]^G$-module (see Section 1.5), so that ϕ restricts to give an isomorphism

$$K[V]^G_+ \otimes_{K[V]^G} K[V]^G \xrightarrow{\cong} K[V]^G_+ \subseteq K[V].$$

This contradiction shows that $\ker(\phi) = 0$, as required.

Finally, we offer two proofs that (ii) $\Rightarrow$ (i). The first only works in coprime characteristic, but has the advantage that it does not use much machinery. The second is effectively Auslander's proof of the purity of the branch locus [8] (see also Nagata [66], Section 41 and Bourbaki [17] Chapter 5 §5, Exercise 7). It works in arbitrary characteristic, but depends on much of the material in Chapters 3 and 6.

The first proof of (ii) $\Rightarrow$ (i) goes as follows. Suppose that

$$K[V]^G = K[\psi_1, \dots, \psi_n]$$

is a polynomial ring with $\deg(\psi_i) = k_i$, arranged so that $k_1 \leq k_2 \leq \dots \leq k_n$. Then the Poincaré series is

$$p(K[V]^G, t) = \prod_{i=1}^{n} \frac{1}{(1-t)^n}.$$

Recall from Section 2.6 that the Laurent expansion of the Poincaré series begins

$$p(K[V]^G, t) = \frac{1}{|G|}\left(\frac{1}{(1-t)^n} + \frac{r/2}{(1-t)^{n-1}} + \cdots\right)$$

where r is the number of pseudoreflections in the action of G on V. Equating these two expressions and multiplying by $|G|(1-t)^n$, we obtain

$$|G| \prod_{i=1}^{n} \frac{1}{1 + t + \dots + t^{k_i - 1}} = 1 + \frac{r}{2}(1-t) + O(1-t)^2.$$

Evaluating at $t = 1$ gives $|G| = k_1 \dots k_n$. On the other hand, we can differentiate both sides to obtain

$$-|G|\left(\prod_{i=1}^{n}\frac{1}{1+t+\cdots+t^{k_i-1}}\right)\left(\sum_{i=1}^{n}\frac{1+2t+\cdots+(k_i-1)t^{k_i-2}}{1+t+\cdots+t^{k_i-1}}\right) = -\frac{r}{2}+O(1-t).$$

Now set $t = 1$ to obtain

$$-\frac{|G|}{k_1\dots k_n}\sum_{i=1}^{n}\frac{k_i(k_i-1)/2}{k_i} = -\frac{r}{2}$$

so that

$$r = \sum_{i=1}^{n}(k_i - 1).$$

In particular, if $G \neq 1$, then some k_i is bigger than one, and so some non-trivial element of G acts on V as a pseudoreflection.

Let H be the subgroup of G generated by the pseudoreflections. Since we have already proved that (i) $\Rightarrow$ (v), we know that $K[V]^H$ is a polynomial ring, say

$$K[V]^H = K[\eta_1, \dots, \eta_n]$$

with $\deg(\eta_i) = l_i$, arranged so that $l_1 \le l_2 \le \cdots \le l_n$. Let $\psi_i = p_i(\eta_1, \dots, \eta_n)$. For each value of i, the polynomials $p_1, \dots, p_i$ cannot only involve $\psi_1, \dots, \psi_{i-1}$, since otherwise $p_1, \dots, p_i, \psi_i, \dots, \psi_n$ would be algebraically independent. So for some values of $m_i \le i \le m_i'$, p_{m_i} involves $\psi_{m_i'}$ and so

$$k_i \ge k_{m_i} \ge l_{m_i'} \ge l_i.$$

Since G and H have the same pseudoreflections, we know that

$$\sum_{i=1}^{n}(k_i - 1) = \sum_{i=1}^{n}(l_i - 1).$$

So each k_i must be equal to l_i, and hence $|G| = |H|$.

Finally, we give the proof of (ii) $\Rightarrow$ (i) which works in arbitrary characteristic. We begin by observing that the fixed points of any group element g form a linear subspace, which has codimension one if and only if g is a pseudoreflection. The branch locus (see Section 3.11) of $K[V]^G \subseteq K[V]$ is the union of the fixed point spaces.

Let H be a subgroup of G generated by pseudoreflections. Let A and B be the localizations of $K[V]^G$ and $K[V]^H$ at the ideals of positive degree elements. Then $A = B^{G/H}$ is a regular local ring, since it is a localization of a polynomial ring, and B is a reflexive A-module since it is integrally closed.

Now form the twisted group ring $B[G/H]$; namely the free B-module on the elements of G/H with multiplication given by $(b_1g_1)(b_2g_2) = (b_1g_1(b_2))(g_1g_2)$. The map $B[G/H] \to \mathrm{Hom}_A(B,B)$ given by sending bg to the map $b' \mapsto bg(b')$ is then a ring homomorphism. If $\mathfrak{P}$ and $\mathfrak{p} = \mathfrak{P} \cap A$ are height one primes in B and A then $B_\mathfrak{P}$ is a finitely generated free $A_\mathfrak{p}$-module and so $B_\mathfrak{P}[G/H] \to \mathrm{Hom}_{A_\mathfrak{p}}(B_\mathfrak{P}, B_\mathfrak{P})$ is an isomorphism. Since $B[G/H]$ and $\mathrm{Hom}_A(B,B)$ are reflexive B-modules, it follows that $B[G/H] \to \mathrm{Hom}_A(B,B)$ is an isomorphism, and so $\mathrm{Hom}_A(B,B)$ is isomorphic to a direct sum of copies of B. So applying Theorem 6.4.3, we see that B is a free A-module. Now applying Theorem 3.11.1 we have $\mathfrak{D}_{B/A} = \mathfrak{H}_{B/A}$. On the other hand, by Theorem 3.10.2 we have $\mathfrak{D}_{B/A} = 1$ since no element of G/H other than the identity fixes a prime ideal in B of height one and acts trivially on the quotient. So $\mathfrak{H}_{B/A} = 1$, which means that the branch locus is empty. However, if $G/H \neq 1$ then the origin is in the branch locus, and so we have $G/H = 1$, which completes the proof. □

7.3 A theorem of Solomon

In this section, we prove a theorem of L. Solomon [98], which examines the invariants of pseudoreflection groups on polynomial tensor exterior algebras. This section depends heavily on Section 5.4.

Theorem 7.3.1 (Solomon) *Suppose that V is a finite dimensional representation of a finite group G over a field K whose characteristic does not divide $|G|$. Suppose that G is generated by pseudoreflections, so that by Theorem 7.2.1 $K[V]^G = K[f_1,\dots,f_n]$ is a polynomial ring. Then*

$$(K[V] \otimes \Lambda(V^*))^G = K[f_1,\dots,f_n] \otimes \Lambda(df_1,\dots,df_n).$$

Proof Let $K[V] = K[x_1,\dots,x_n]$, so that $K(V) = K(x_1,\dots,x_n)$. By Proposition 1.1.1, $K(V)$ is a Galois extension of $K(f_1,\dots,f_n)$. In particular, it is separable, and so by Proposition 5.4.2, the Jacobian $\mathfrak{j} = \det(\partial f_i/\partial x_j)$ is non-zero.

Recall from Section 5.4 that $\mathfrak{j}$ is a relative invariant for the determinant representation. We claim that in fact

$$K[V]^G_{\det} = \mathfrak{j}K[V]^G.$$

To prove this, we can either notice from Watanabe's Theorem 4.6.2 that $K[V]^G_{\det} = \omega(K[V]^G)$ is a free module on a single generator in the same degree as $\mathfrak{j}$, or we can identify the relative invariants more explicitly as follows.

If $g \in G$ is a pseudoreflection, then we write $\alpha_g \in K[V]$ for a degree two element (recall that we are regarding the generators x_i as degree two elements) which vanishes on the reflecting hyperplane V^g. We claim that if $f \in K[V]^G_{\det}$ and g is a pseudoreflection of order r then f is divisible by α_g^{r-1}. To see this, let λ be the non-trivial eigenvalue of g, and choose a basis $x_1, \ldots, x_n$ for V^* in such a way that g fixes $x_1, \ldots, x_{n-1}$ and $g(x_n) = \lambda^{-1}x_n$, so that we may take $\alpha_g = x_n$. Since $f \in K[V]^G_{\det}$ we have

$$f(x_1, \ldots, x_{n-1}, \lambda^{-1}x_n) = g(f)(x_1, \ldots, x_n) = \lambda f(x_1, \ldots, x_n).$$

Using the chain rule, we differentiate q times to obtain

$$\lambda^{-q}\frac{\partial^q f}{\partial x_n^q}(x_1, \ldots, x_{n-1}, \lambda^{-1}x_n) = \lambda\frac{\partial^q f}{\partial x_n^q}(x_1, \ldots, x_n).$$

For $0 \leq q \leq r-2$, we have $\lambda^{-q} \neq \lambda$, and so $(\partial^q f/\partial x_n^q)(x_1, \ldots, x_{n-1}, 0) = 0$ so that f is divisible by x_n^{r-1} as required.

Now by Theorem 7.2.1, deg($\mathfrak{j}$) is equal to twice the number of pseudoreflections, and so the above argument shows that $\mathfrak{j}$ is a scalar multiple of the product of all the α_g^{r-1}, and so $\mathfrak{j}$ divides every element of $K[V]^G_{\det}$.

Now let G act on

$$K(V) \otimes \Lambda(V^*) = K(x_1, \ldots, x_n) \otimes \Lambda(dx_1, \cdots, dx_n).$$

We first prove that the $\binom{n}{j}$ differential forms $df_{i_1} \wedge \ldots \wedge df_{i_j}$ are linearly independent over $K(V)$. If there is a linear relation

$$\sum_{\{i_1, \ldots, i_j\}} a_{i_1 \ldots i_j} df_{i_1} \wedge \ldots \wedge df_{i_j} = 0$$

then for each subset $\{i_1, \ldots, i_j\}$, we multiply by the remaining df_l with l not in $\{i_1, \ldots, i_j\}$ to get

$$0 = \pm a_{i_1,\ldots,i_j} df_1 \wedge \ldots \wedge df_n = \pm a_{i_1,\ldots,i_j}\, \mathfrak{j}\, dx_1 \wedge \ldots \wedge dx_n.$$

But $\mathfrak{j} \neq 0$, and so $a_{i_1,\ldots,i_j} = 0$.

Since $K(V) \otimes \Lambda^j(V^*)$ has dimension $\binom{n}{j}$ as a vector space over $K(V)$, this proves that it is spanned by the $df_{i_1} \wedge \ldots \wedge df_{i_j}$.

If $\omega = \sum a_{i_1,\ldots,i_j} df_{i_1} \wedge \ldots \wedge df_{i_j} \in K[V] \otimes \Lambda^j(V^*)$ is invariant, then each $a_{i_1,\ldots,i_j}$ is an element of $K(V)^G$ because $df_{i_1} \wedge \cdots \wedge df_{i_j}$ is invariant. Multiplying as before by the remaining df_l with $l \notin \{i_1, \ldots, i_j\}$, we see that

$$a_{i_1,\ldots,i_j} df_1 \wedge \ldots \wedge df_n = a_{i_1,\ldots,i_j}\, \mathfrak{j}\, dx_1 \wedge \ldots \wedge dx_n$$

is also invariant, so $a_{i_1,\ldots,i_j}\, \mathfrak{j} \in K[V]^G_{\det}$. We showed above that $K[V]^G_{\det} = \mathfrak{j}\, K[V]^G$, so it follows that $a_{i_1,\ldots,i_j} \in K[V]^G$, which completes the proof of the theorem. □

Remark Solomon's theorem is false in non-coprime characteristic. For example, if $G = SL_n(\mathsf{F}_q)$, $n \geq 2$, $q \geq 3$, and V is the natural module of dimension n over a field K containing F_q, then $K[V]^G$ is a polynomial ring (see Section 8.2) but $(K[V] \otimes \Lambda(V^*))^G$ contains elements not in the subring generated by these polynomials and their differentials. The problem is that $K[V]^G_{\det}$ is bigger than $\mathrm{j}K[V]^G$, so that there are too many invariants in the top exterior degree. The rest of Solomon's argument works in arbitrary characteristic, and so the conclusion of the theorem holds if and only if $K[V]^G_{\det} = \mathrm{j}K[V]^G$.

Chapter 8

Modular invariants

8.1 Dickson's theorem

Let q be a power of a prime p, and let E be the natural n dimensional representation of the finite general linear group $GL_n(\mathsf{F}_q)$ over F_q. If K is a field containing F_q, let $V = K \otimes_{\mathsf{F}_q} E$. Dickson [31] proved that the ring of invariants $K[V]^{GL_n(\mathsf{F}_q)}$ is a polynomial ring $K[c_0, \dots, c_{n-1}]$ on generators c_j in degree $q^n - q^j$. Dickson's proof has been cleaned up considerably (Wilkerson [112]). We present a variation due to Tamagawa of the proof given in [112], based on a lecture of Larry Smith in Göttingen. Note that extending the field K does not affect whether the invariants form a polynomial ring, and so we may assume without loss of generality that K is infinite.

We begin, if you like, at the wrong end, by letting L_0 be the field of fractions of the polynomial ring $K[c_0, \dots, c_{n-1}]$ in some indeterminates c_j which, when appropriate, will be regarded as being in degree $q^n - q^j$. Consider the polynomial

$$f(X) = X^{q^n} - c_{n-1}X^{q^{n-1}} + c_{n-2}X^{q^{n-2}} - \cdots + (-1)^n c_0 X \in L_0[X]$$

and let L be a splitting field of $f(X)$ over L_0. Note that if we set $\deg(X) = 1$ then the above polynomial is homogeneous of degree q^n.

The basic property we shall use of the polynomial $f(X)$ is that it is F_q-linear. Namely, if $\lambda, \mu \in \mathsf{F}_q$ then

$$f(\lambda X + \mu Y) = \lambda f(X) + \mu f(Y).$$

It follows that the zeros of f form an F_q-vector subspace W of L. Since the derivative $f'(X) = (-1)^n c_0 \neq 0$, f has no repeated roots, and so there are q^n zeros, which means that $\dim_{\mathsf{F}_q}(W) = n$. In particular L is a Galois extension of L_0.

We have

$$\prod_{w \in W} (X - w) = X^{q^n} - c_{n-1}X^{q^{n-1}} + c_{n-2}X^{q^{n-2}} - \cdots + (-1)^n c_0 X$$

and so the c_j are expressible as elementary symmetric polynomials in the elements $w \in W$. So if we set $E = W^* = \mathrm{Hom}_{\mathbb{F}_q}(W, \mathbb{F}_q)$ and $V = K \otimes_{\mathbb{F}_q} E = \mathrm{Hom}_{\mathbb{F}_q}(W, K)$, then L is generated over K by the elements $w \in V^*$. Since the c_j are algebraically independent, the transcendence degree of L over K is n, and so $L = K(V)$ is the field of fractions of the ring $K[V]$ of polynomial functions on V.

The action of $GL_n(\mathbb{F}_q)$ on W gives rise to an action on E and V, and hence on $K[V]$ and L. The elements $c_0, \dots, c_{n-1}$ are invariant under the action, and so L_0 is contained in the fixed field. So we have $GL_n(\mathbb{F}_q) \leq \mathrm{Gal}(L/L_0)$. On the other hand, a Galois automorphism of L fixing L_0 is determined by its effect on the roots of $f(X)$, and this action in particular has to be $\mathbb{F}_q$-linear. So $\mathrm{Gal}(L/L_0) = GL_n(\mathbb{F}_q)$, and hence $L^{GL_n(\mathbb{F}_q)} = L_0$.

Finally, we have (cf. Proposition 1.1.1)

$$K[V]^{GL_n(\mathbb{F}_q)} = L^{GL_n(\mathbb{F}_q)} \cap K[V] = K(c_0, \dots, c_{n-1}) \cap K[V].$$

Now $K[V]$, and hence also $K[V]^{GL_n(\mathbb{F}_q)}$, is integral over $K[c_0, \dots, c_{n-1}]$. Since the latter is integrally closed, we deduce that

$$K[V]^{GL_n(\mathbb{F}_q)} = K[c_0, \dots, c_{n-1}].$$

To summarize, we have proved the following theorem.

Theorem 8.1.1 (Dickson) *Let $G = GL_n(\mathbb{F}_q)$, and let K be a field containing $\mathbb{F}_q$. Let E be the natural module for G over $\mathbb{F}_q$, and let $V = K \otimes_{\mathbb{F}_q} E$ be the natural module over K. Then*

$$K[V]^G = K[c_0, \dots, c_{n-1}],$$

where the c_j are defined by the equation

$$\prod_{w \in E^*} (X - w) = X^{q^n} - c_{n-1}X^{q^{n-1}} + c_{n-2}X^{q^{n-2}} - \cdots + (-1)^n c_0 X.$$

□

If we wish to make the dimension n of V explicit in the notation for the Dickson invariants, we write $c_{n,j}$ for what we have been calling c_j. The following lemma describes what happens when we vary n.

Lemma 8.1.2 *Suppose that E' is an $\mathbb{F}_q$-subspace of E of codimension m. Then the restriction of $c_{n,j}$ to a function on $V' = K \otimes_{\mathbb{F}_q} E'$ is $c_{n-m,j-m}^{q^m}$ if $m \leq j$ and zero if $m > j$.*

Proof The restriction of $\prod_{w \in E^*} (X - w)$ to E' is $\prod_{w \in (E')^*} (X - w)^{q^m}$. □

Proposition 8.1.3 (Ian Macdonald) *The Dickson invariant c_m is equal to the sum over all subspaces $E' \leq E$ of codimension m, of the product of the $q^n - q^m$ linear forms in E^* which have non-zero value on E':*

$$c_m = \sum_{\dim_{\mathbb{F}_q}(E')=n-m} \prod_{w|_{E'}\neq 0} w.$$

Proof The right hand side is clearly an invariant of the same degree as c_m, so it must be a multiple of c_m by the above theorem. To evaluate the coefficient, we choose a particular subspace E' of codimension m, and restrict both sides to E'. The left hand side $c_{n,m}$ restricts to $c_{n-m,0}^{q^m}$ by the lemma. All the terms in the sum on the right hand side vanish except the one corresponding to the chosen E', and this term restricts to $c_{n-m,0}^{q^m}$. Therefore the coefficient is one, as required. □

8.2 The special linear group

When we restrict down to $SL_n(\mathbb{F}_q)$, we see that the invariant $(-1)^n c_0$ has a $(q-1)$st root, as follows. We choose one non-zero element w from each line in E^*, and let u be the product of these w's. Since the product of the non-zero elements of $\mathbb{F}_q$ is equal to -1, and the number $(q^n - 1)/(q - 1) = q^{n-1} + \cdots + 1$ of lines is congruent to n mod two for q odd, we have $u^{q-1} = (-1)^n c_0$. An element $g \in GL_n(\mathbb{F}_q)$ sends u to some multiple $\lambda(g)u$, and $\lambda : GL_n(\mathbb{F}_q) \to \mathbb{F}_q^\times$ is a linear character, hence equal to some power of the determinant. So u is invariant under $SL_n(\mathbb{F}_q)$. Note that u is only well defined up to multiplication by scalars, since it is not $GL_n(\mathbb{F}_q)$-invariant. As with the Dickson invariants, if we wish to make the dimension n of V explicit, we write u_n rather than u.

Theorem 8.2.1 *Let $G = SL_n(\mathbb{F}_q)$, and let K be a field containing $\mathbb{F}_q$. Let E be the natural module for G over $\mathbb{F}_q$, and let $V = K \otimes_{\mathbb{F}_q} E$ be the natural module over K. Then*

$$K[V]^G = K[u, c_1, \ldots, c_{n-1}],$$

where the c_j are defined as in Theorem 8.1.1, and u is defined above.

Proof The field of fractions $K(V)$ is a Galois extension of $K(u, c_1, \ldots, c_{n-1})$ whose Galois group lies between $SL_n(\mathbb{F}_q)$ and $GL_n(\mathbb{F}_q)$. Computing the degree of the extension using Poincaré series (see Section 2.4), we see that the Galois group must be $SL_n(\mathbb{F}_q)$. Thus we have

$$K[V]^{SL_n(\mathbb{F}_q)} = K(V)^{SL_n(\mathbb{F}_q)} \cap K[V] = K(u, c_1, \ldots, c_{n-1}) \cap K[V].$$

Since $K[V]$, and hence $K[V]^{SL_n(\mathbb{F}_q)}$, is integral over $K[c_0, \ldots, c_{n-1}]$, and hence also over $K[u, c_1, \ldots, c_{n-1}]$, and the latter is integrally closed, we deduce that

$$K[V]^{SL_n(\mathbb{F}_q)} = K[u, c_1, \ldots, c_{n-1}],$$

as required. □

8.3 Symplectic invariants

Let q be a power of a prime p, and let E be a vector space over $\mathbb{F}_q$ of dimension $2n$, say with basis $e_1, \ldots, e_{2n}$. We endow E with a non-singular symplectic form $\langle -, - \rangle$ given (without loss of generality) by

$$\langle \sum \lambda_i e_i, \sum \mu_i e_i \rangle = \lambda_1\mu_2 - \lambda_2\mu_1 + \lambda_3\mu_4 - \lambda_4\mu_3 + \cdots$$

Let K be an infinite field containing $\mathbb{F}_q$, and set $V = K \otimes_{\mathbb{F}_q} E$, so that the symplectic form extends in an obvious way to V. Let G be the finite symplectic group $Sp_{2n}(\mathbb{F}_q)$, for which E is the natural module over $\mathbb{F}_q$ and V is the natural module over K. We describe in this section the calculation by Carlisle and Kropholler [21] of the invariants $K[V]^G$. The answer is not a polynomial ring, but is an example of a **complete intersection.** This means that it has a presentation in which the number of generators minus the number of relations is equal to the Krull dimension (i.e., $2n$).

Denote by F the **Frobenius morphism** of V given by $F(\sum \lambda_i e_i) = \sum \lambda_i^q e_i$, so that E is the fixed points of F on V. Then as well as the Dickson invariants, there are some obvious invariants ξ_i of degree $q^i + 1$ given by $\xi_i(v) = \langle v, F^i(v) \rangle$. If $x_1, \ldots, x_{2n}$ is the basis of V^* dual to $e_1, \ldots, e_{2n}$, then

$$\xi_i = x_1 x_2^{q^i} - x_2 x_1^{q^i} + x_3 x_4^{q^i} - x_4 x_3^{q^i} + \cdots$$

The presentation of $K[V]^G$ given by Carlisle and Kropholler consists of the $3n-1$ generators $c_n, \ldots, c_{2n-1}, \xi_1, \ldots, \xi_{2n-1}$, subject to $n-1$ relations (Theorem 8.3.11).

Lemma 8.3.1 *$\xi_1, \ldots, \xi_{2n}$ are algebraically independent, and $K(V)$ is a finite separable extension of $K(\xi_1, \ldots, \xi_{2n})$.*

Proof This follows from the Jacobian criterion (Proposition 5.4.2), since

$$\det\left(\frac{\partial \xi_i}{\partial x_j}\right) = \begin{vmatrix} x_2^q & -x_1^q & \cdots \\ x_2^{q^2} & -x_1^{q^2} & \\ \vdots & & \\ x_2^{q^{2n}} & -x_1^{q^{2n}} & \cdots \end{vmatrix} \neq 0.$$

□

Lemma 8.3.2 *If $v_0, \ldots, v_{2n}$ are elements of V and $\lambda : V \to k$ is a linear form, then*

$$\sum_\sigma \operatorname{sgn}(\sigma) \langle v_{\sigma(0)}, v_{\sigma(1)} \rangle \cdots \langle v_{\sigma(2n-2)}, v_{\sigma(2n-1)} \rangle \lambda(v_{\sigma(2n)}) = 0.$$

Here, σ runs over a set of coset representatives of the wreath product $\Sigma_2 \wr \Sigma_n$ in Σ_{2n+1}. The subgroup $\Sigma_2 \wr \Sigma_n$ is the centralizer of the permutation

$$(0,1)(2,3)\cdots(2n-2,2n-1)(2n).$$

Since $\langle x,y\rangle = -\langle y,x\rangle$, the expression inside the above summation only depends on the coset representative of $\Sigma_2 \wr \Sigma_n$ in Σ_{2n+1} given by σ.

Proof The sum is a totally alternating multilinear function in $2n+1$ variables taken from the $2n$ dimensional vector space V, and must hence be zero. □

Proposition 8.3.3 *We have*

$$\sum_{\sigma} \mathrm{sgn}(\sigma)\xi^{q^{\sigma(0)}}_{\sigma(1)-\sigma(0)}\cdots\xi^{q^{\sigma(2n-2)}}_{\sigma(2n-1)-\sigma(2n-2)}X^{q^{\sigma(2n)}} = u_{2n}\prod_{w\in E^*}(X-w).$$

Again, σ runs through a set of coset representatives of $\Sigma_2 \wr \Sigma_n$ in Σ_{2n+1}, but we must choose the coset representatives in such a way that each $\sigma(2j-1)-\sigma(2j-2)$ is positive. Note that u_{2n} is the invariant of the special linear group introduced in the last section, and is only well defined up to scalar multiplication, but this proposition gives us a particular choice of representative, which we use from now on.

Proof The left hand side vanishes whenever X is replaced by $\lambda \in V^*$, by the lemma. It also vanishes when restricted to proper subspaces defined over $\mathbb{F}_q$, and so it is divisible by the right hand side. Furthermore, the ξ_j are algebraically independent, so the left hand side is non-zero. Now compare degrees. □

It follows by comparing the above proposition with Theorem 8.1.1 that there are polynomials $D_{2n,j}$ $(0 \le j \le 2n)$ with

$$D_{2n,j}(\xi_1,\dots,\xi_{2n}) = u_{2n}c_{2n,j}.$$

Here, we have used the obvious convention that $c_{2n,2n} = 1$. We set $U_{2n} = D_{2n,2n}$, which does not involve ξ_{2n}, so that

$$U_{2n}(\xi_1,\dots,\xi_{2n-1}) = u_{2n}.$$

Note also that $D_{2n,0} = U^q_{2n}$, since $c_{2n,0} = u^{q-1}_{2n}$.

Theorem 8.3.4 *We have* $K[V]^{Sp_{2n}(\mathbb{F}_q)}[c_0^{-1}] = K[\xi_1,\dots,\xi_{2n}][U_{2n}^{-1}]$.

Proof We have field extensions

$$K(x_1,\dots,x_{2n}) \ge K(\xi_1,\dots,\xi_{2n}) \ge K(c_{2n,0},\dots,c_{2n,2n-1}).$$

By Theorem 8.1.1, the left hand side is a Galois extension of the right hand side with Galois group $GL_{2n}(\mathbb{F}_q)$. The stabilizer of ξ_1 in $GL_{2n}(\mathbb{F}_q)$ is $Sp_{2n}(\mathbb{F}_q)$, and this stabilizes $\xi_2, \dots, \xi_{2n}$. So the left hand field extension is Galois with group $Sp_{2n}(\mathbb{F}_q)$. So $K(V)^{Sp_{2n}(\mathbb{F}_q)} = K(\xi_1, \dots, \xi_{2n})$.

Now the ring $K[\xi_1, \dots, \xi_{2n}][U_{2n}^{-1}]$ is integrally closed and contains the $c_{2n,j}$. So $K[V]^{Sp_{2n}(\mathbb{F}_q)}[c_0^{-1}]$ is integral over it, and is contained in its field of fractions, hence equal to it. □

Warning The ring $K[x_1, \dots, x_{2n}]$ is not a finite extension of $K[\xi_1, \dots, \xi_{2n}]$, so we cannot use the method of Section 2.4 to calculate the degree of the field extension at this stage.

Lemma 8.3.5 *The polynomial $U_{2n} \in K[\xi_1, \dots, \xi_{2n-1}]$ is irreducible, and the kernel of the composite map*

$$K[\xi_1, \dots, \xi_{2n-1}] \hookrightarrow K[\xi_1, \dots, \xi_{2n}] \hookrightarrow K[x_1, \dots, x_{2n}] \twoheadrightarrow K[x_1, \dots, x_{2n-2}]$$

(where the last map is given by setting $x_{2n-1} = x_{2n} = 0$) is the principal ideal generated by U_{2n}.

Proof The image of U_{2n} in $K[x_1, \dots, x_{2n}]$ is u_{2n}, which is a product of degree one terms corresponding to the lines in E^*. These are permuted transitively by $Sp_{2n}(\mathbb{F}_q)$, so u_{2n} is irreducible as an element of $K[V]^{Sp_{2n}(\mathbb{F}_q)}$. It follows that U_{2n} is irreducible as an element of $K[\xi_1, \dots, \xi_{2n-1}]$, so that by Theorem 2.3.2, $K[\xi_1, \dots, \xi_{2n-1}]/(U_{2n})$ is an integral domain of Krull dimension $2n-2$ and every proper quotient has smaller Krull dimension. By Lemma 8.3.1, the images of $\xi_1, \dots, \xi_{2n-2}$ in $K[x_1, \dots, x_{2n-2}]$ are algebraically independent, so there can be no further kernel. □

Lemma 8.3.6 *For $1 \le j \le 2n-1$, as a polynomial in $\xi_1, \dots, \xi_{2n}$, $D_{2n,j}$ is of the form $D^q_{2n-2,j-1}\xi_{2n}$ plus a polynomial in $\xi_1, \dots, \xi_{2n-1}$.*

Proof This follows from the form of the equation given in Proposition 8.3.3. □

Proposition 8.3.7 *There are polynomials $P_{i,j} \in K[\xi_1, \dots, \xi_{2j-1}]$ $(1 \le i \le j)$ which are independent of n (regarding the ξ's as formal variables), and such that for $1 \le i \le n$,*

$$D_{2n,n-i} = \sum_{j=i}^{n} P_{i,j}^{q^{n-j}} D_{2n,n+j}.$$

As a function of $x_1, \dots, x_{2n}$, $P_{i,j}$ has degree $q^{2j} - q^{j-i}$.

Proof For a given value of i, we prove this by induction on $n \ge i$. If $n = i$ then there is only one term in the sum, with $j = n$, and we set $P_{n,n} = U_{2n}^{q-1}$. The equation then says that $D_{2n,0} = U_{2n}^{q-1} D_{2n,2n}$, which is true since both sides are equal to U_{2n}^q. So we assume that $n > i$ and that the proposition is true for $n-1$. In particular, $P_{i,j}$ has been defined for $i \le j < n$, and we only need to define $P_{i,n}$. So we examine the polynomial

$$\Phi_{i,n} = D_{2n,n-i} - \sum_{j=i}^{n-1} P_{i,j}^{q^{n-j}} D_{2n,n+j} \in K[\xi_1, \ldots, \xi_{2n}].$$

By Lemma 8.3.6, we can write this as

$$\left(D_{2n-2,n-i-1} - \sum_{j=i}^{n-1} P_{i,j}^{q^{n-j-1}} D_{2n-2,n-j-1} \right)^q \xi_{2n}$$

plus a polynomial in $\xi_1, \ldots, \xi_{2n-1}$. By the inductive hypothesis, the coefficient of ξ_{2n} vanishes, and so $\Phi_{i,n} \in K[\xi_1, \ldots, \xi_{2n-1}]$. Its image under the composite map

$$K[\xi_1, \ldots, \xi_{2n-1}] \hookrightarrow K[\xi_1, \ldots, \xi_{2n}] \hookrightarrow K[x_1, \ldots, x_{2n}] \twoheadrightarrow K[x_1, \ldots, x_{2n-2}]$$

is zero, because each $D_{2n,j}$ goes to $u_{2n} c_{2n,j}$ in $K[x_1, \ldots, x_{2n}]$ and then to zero in $K[x_1, \ldots, x_{2n-2}]$. So by Lemma 8.3.5, $\Phi_{i,n}$ is divisible by $D_{2n,2n} = U_{2n}$ in $K[\xi_1, \ldots, \xi_{2n-1}]$, and we set $P_{i,n}$ equal to the quotient. □

Lemma 8.3.8 *If $v_0, \ldots, v_{2n}$ are elements of V then*

$$\sum_{j=0}^{2n} (-1)^j \langle v_0, v_j \rangle \det(v_0, \ldots, v_{j-1}, v_{j+1}, \ldots, v_{2n}) = 0.$$

Proof This expression is a totally alternating multilinear function of $2n+1$ variables taken from the $2n$ dimensional vector space V, and must hence be zero. □

Proposition 8.3.9 *For $0 \le i \le 2n-1$ we have*

$$\sum_{j=0}^{i-1} (-1)^j \xi_{i-j}^{q^j} c_{2n,j} - \sum_{j=i+1}^{2n} (-1)^j \xi_{j-i}^{q^i} c_{2n,j} = 0.$$

Proof The value of this on a vector $v \in V$ is

$$\sum_{j=0}^{2n} (-1)^j \langle F^i(v), F^j(v) \rangle \det(v, \ldots, F^{j-1}(v), F^{j+1}(v), \ldots, F^{2n}(v))$$

which is zero by Lemma 8.3.8. □

In particular with $i = 0$ we obtain the following.

Corollary 8.3.10 *We have* $\xi_{2n} - c_{2n,2n-1}\xi_{2n-1} + \cdots + c_{2n,1}\xi_1 = 0$. □

Theorem 8.3.11 *The ring* $K[V]^{Sp_{2n}(\mathbb{F}_q)}$ *is generated by the elements*

$$c_0, \ldots, c_{2n-1}, \xi_1, \ldots, \xi_{2n}$$

subject only to the following relations:

(a) $\displaystyle\sum_{j=0}^{i-1}(-1)^j \xi_{i-j}^{q^j} c_j = \sum_{j=i+1}^{2n}(-1)^j \xi_{j-i}^{q^i} c_j \;(0 \leq i \leq n-1)$

(b) $\displaystyle c_i = \sum_{j=n-i}^{n} P_{n-i,j}(\xi_1, \ldots, \xi_{2j-1})^{q^{n-j}} c_{n+j} \;(0 \leq i \leq n-1)$

Note that (b) and the case $i = 0$ of (a) can be used to eliminate $c_0, \ldots, c_{n-1}$ and ξ_{2n}, giving a presentation with $3n-1$ generators $c_n, \ldots, c_{2n-1}, \xi_1, \ldots, \xi_{2n-1}$ and $n-1$ relations (the cases $1 \leq i \leq n-1$ of (a) after the substitutions have been made).

Proof We introduce formal variables $C_0, \ldots, C_{2n-1}, Y_1, \ldots, Y_{2n}$, and define a map

$$S = K[C_0, \ldots, C_{2n-1}, Y_1, \ldots, Y_{2n}] \to K[V]^{Sp_{2n}(\mathbb{F}_q)}$$

by sending C_j to c_j and Y_j to ξ_j (recall our convention that $c_{2n} = 1$). Write $R_0, \ldots, R_{n-1}$ for the elements of S corresponding to the left hand side minus the right hand side of (a), and $R'_0, \ldots, R'_{n-1}$ for the elements corresponding to the left hand side minus the right hand side of (b). We write the relations R and R' in matrix form as follows:

$$\begin{pmatrix}
0 & Y_1 & -Y_2 & \cdots & \mp Y_{n-1} & \pm Y_n & \mp Y_{n+1} & \cdots & Y_{2n-1} \\
-Y_1 & 0 & -Y_1^q & & \mp Y_{n-2}^q & \pm Y_{n-1}^q & \mp Y_n^q & & Y_{2n-2}^q \\
-Y_2 & Y_1^q & 0 & & \mp Y_{n-3}^{q^2} & \pm Y_{n-2}^{q^2} & \mp Y_{n-1}^{q^2} & & Y_{2n-3}^{q^2} \\
\vdots & & \ddots & & & & & & \vdots \\
-Y_{n-1} & Y_{n-2}^q & -Y_{n-3}^{q^2} & \cdots & 0 & \pm Y_1^{q^{n-1}} & \mp Y_2^{q^{n-1}} & \cdots & Y_n^{q^{n-1}} \\
-1 & 0 & 0 & \cdots & 0 & 0 & 0 & \cdots & 0 \\
0 & -1 & 0 & & 0 & 0 & 0 & 0 & P_{n-1,n-1}^q \\
0 & 0 & -1 & & 0 & 0 & 0 & P_{n-2,n-2}^{q^2} & P_{n-2,n-1}^q \\
\vdots & & & & & & & & \vdots \\
0 & 0 & 0 & \cdots & -1 & 0 & P_{1,1}^{q^{n-1}} & \cdots & P_{1,n-1}^q
\end{pmatrix}
\begin{pmatrix}
C_0 \\ C_1 \\ C_2 \\ \vdots \\ C_{n-1} \\ C_n \\ C_{n+1} \\ C_{n+2} \\ \vdots \\ C_{2n-1}
\end{pmatrix}
=
\begin{pmatrix}
Y_{2n} \\ Y_{2n-1}^q \\ Y_{2n-2}^{q^2} \\ \vdots \\ Y_{n+1}^{q^{n-1}} \\ -P_{n,n} \\ -P_{n-1,n} \\ -P_{n-2,n} \\ \vdots \\ -P_{1,n}
\end{pmatrix}$$

By Propositions 8.3.7 and 8.3.9 we have a map

$$\bar{S} = S/(R_0, \ldots, R_{n-1}, R'_0, \ldots, R'_{n-1}) \to K[V]^{Sp_{2n}(\mathbb{F}_q)}$$

which we claim is an isomorphism.

We first note that

$$Y_1, \dots, Y_n, R_{n-1}, \dots, R_0, C_{2n-1}, \dots, C_n, R'_{n-1}, \dots, R'_0$$

is a regular sequence, because

$$\begin{aligned} R_j &\equiv Y_{2n-j}^{q^j} \bmod (Y_1, \dots, Y_{2n-j-1}) \\ R'_j &\equiv C_j \bmod (Y_1, \dots, Y_{2n}, C_{2n-1}, \dots, C_n). \end{aligned}$$

Write δ for the determinant of the above matrix Δ. Then

$$\delta \equiv Y_n^{(q^n-1)/(q-1)} \bmod (Y_1, \dots, Y_{n-1})$$

and so $R_0, \dots, R_{n-1}, R'_0, \dots, R'_{n-1}, \delta$ is also a regular sequence in S. Thus δ is not a zero divisor in $\bar{S}$ and so $\bar{S}$ embeds in $\bar{S}[\delta^{-1}]$. Moreover, in $\bar{S}[\delta^{-1}]$, we can express $C_0, \dots, C_{2n-1}$ in terms of $Y_1, \dots, Y_{2n}$, so that

$$\bar{S}[\delta^{-1}] = K[Y_1, \dots, Y_{2n}][\delta^{-1}]$$

is an integral domain, and hence so is $\bar{S}$.

Next, we claim that $\bar{S}$ is a unique factorization domain. By Lemma 6.3.1, since $\bar{S}[\delta^{-1}]$ is a unique factorization domain, it suffices to prove that δ is prime in $\bar{S}$, or equivalently that $\bar{S}/(\delta)$ is an integral domain. Let δ_1 be the determinant of the matrix Δ_1 obtained from Δ by deleting the first row and the last column. Then

$$\delta_1 \equiv Y_{n-1}^{(q^n-q)/(q-1)} \bmod (Y_1, \dots, Y_{n-2})$$

and so $R_0, \dots, R_{n-1}, R'_0, \dots, R'_{n-1}, \delta_1, \delta$ is a regular sequence in S. So $\bar{S}/(\delta)$ embeds in $\bar{S}[\delta_1^{-1}]/(\delta)$, and it suffices to prove that the latter is an integral domain. Since R_0 expresses Y_{2n} in terms of the remaining variables, we may eliminate Y_{2n} from the generators of $\bar{S}[\delta_1^{-1}]/(\delta)$. The relations $R_1, \dots, R_{n-1}, R'_0, \dots, R'_{n-1}$ take the form

$$\Delta_1 \begin{pmatrix} C_0 \\ \vdots \\ C_{2n-2} \end{pmatrix} + \begin{pmatrix} Y_{2n-2}^q \\ \vdots \\ P_{1,n-1}^q \end{pmatrix} C_{2n-1} = \begin{pmatrix} Y_{2n-1}^q \\ \vdots \\ -P_{1,n} \end{pmatrix}$$

and since Δ_1 is invertible in $\bar{S}[\delta_1^{-1}]/(\delta)$, we can solve to express $C_0, \dots, C_{2n-2}$ in terms of $C_{2n-1}, Y_1, \dots, Y_{2n-1}$. Furthermore, $\delta = \delta_1 Y_{2n-1}$ plus a polynomial in $Y_1, \dots, Y_{2n-2}$, so we may eliminate Y_{2n-1}. The remaining variables $Y_1, \dots, Y_{2n-2}$ and C_{2n-1} are independent, and so

$$\bar{S}[\delta_1^{-1}]/(\delta) = K[Y_1, \dots, Y_{2n-2}, C_{2n-1}][\delta_1^{-1}]$$

which is an integral domain.

This completes the proof that $\bar{S}$ is a unique factorization domain, and hence an integrally closed domain. Now the image of $\bar{S}$ in $K[V]$ contains the Dickson invariants, and so $K[V]^{Sp_{2n}(\mathbb{F}_q)}$ is integral over the image. Since the Krull dimension of $\bar{S}$ is $2n$, there can be no kernel, so the image is integrally closed. $K[V]^{Sp_{2n}(\mathbb{F}_q)}$ is contained in the field of fractions of the image, and hence equal to the image. □

Appendix A

Examples over the complex numbers

Denote by ζ_n the primitive nth root of unity $e^{2\pi i/n}$ in $\mathbf{C}$.

(i) $G \cong \mathbf{Z}/n$ generated by the matrix (ζ_n).

$$\mathbf{C}[x]^{\mathbf{Z}/n} = \mathbf{C}[x^n]$$

Poincaré series $1/(1-t^n)$.

(ii) $G \cong \mathbf{Z}/2$ generated by the matrix $\begin{pmatrix} -1 & 0 \\ 0 & -1 \end{pmatrix}$.

Generators for the invariants:

$$\begin{aligned} y_1 &= x_1^2 \\ y_2 &= x_1 x_2 \\ y_3 &= x_2^2. \end{aligned}$$

Relation: $y_1 y_3 = y_2^2$.

Poincaré series $(1+t^2)/(1-t^2)^2$.

(iii) $G \cong D_{2n}$ generated by the matrices $\begin{pmatrix} \zeta_n & 0 \\ 0 & \zeta_n^{-1} \end{pmatrix}$ and $\begin{pmatrix} 0 & 1 \\ 1 & 0 \end{pmatrix}$.

Generators for the invariants:

$$\begin{aligned} y_1 &= x_1 x_2 \\ y_2 &= x_1^n + x_2^n. \end{aligned}$$

(No relations)

Poincaré series $1/(1-t^2)(1-t^n)$.

(iv) $G \cong Q_{4n}$ generated by the matrices $\begin{pmatrix} \zeta_{2n} & 0 \\ 0 & \zeta_{2n}^{-1} \end{pmatrix}$ and $\begin{pmatrix} 0 & i \\ i & 0 \end{pmatrix}$.

Generators for the invariants:

$$\begin{aligned} y_1 &= x_1^2 x_2^2 \\ y_2 &= x_1^{2n} + (-1)^n x_2^{2n} \\ y_3 &= x_1 x_2 (x_1^{2n} - (-1)^n x_2^{2n}). \end{aligned}$$

Relation: $y_3^2 = y_1 y_2^2 + 4(-y_1)^{n+1}$.

Poincaré series $(1+t^{2n+2})/(1-t^4)(1-t^{2n})$.

(v) $G \cong SL_2(\mathsf{F}_3)$ generated by the matrices $\begin{pmatrix} i & 0 \\ 0 & -i \end{pmatrix}$, $\begin{pmatrix} 0 & i \\ i & 0 \end{pmatrix}$ and $\frac{1}{2}\begin{pmatrix} i-1 & i-1 \\ i+1 & -i-1 \end{pmatrix}$.

Generators for the invariants:

$$\begin{aligned} y_1 &= x_1 x_2 (x_1^4 - x_2^4) \\ y_2 &= (x_1^4 + 2i\sqrt{3}x_1^2 x_2^2 + x_2^4)(x_1^4 - 2i\sqrt{3}x_1^2 x_2^2 + x_2^4) \\ y_3 &= (x_1^4 + 2i\sqrt{3}x_1^2 x_2^2 + x_2^4)^3. \end{aligned}$$

Relation: $y_3(y_3 - 12i\sqrt{3}y_1^2) = y_2^3$.

Poincaré series $(1+t^{12})/(1-t^6)(1-t^8)$.

For further examples over the complex numbers and the p-adics, see the groups generated by pseudoreflections discussed in Section 7.1. Two groups related to examples 23 and 24 of Table 7.1 are described explicitly in Burnside [19], §§266–7. For convenience, we present his answers here.

(vi) $G \cong A_5$ generated by the matrices

$$\begin{pmatrix} 1 & 0 & 0 \\ 0 & \zeta & 0 \\ 0 & 0 & \zeta^{-1} \end{pmatrix} \text{ and } \begin{pmatrix} 1 & 1 & 1 \\ 2 & \zeta^2+\zeta^{-2} & \zeta+\zeta^{-1} \\ 2 & \zeta+\zeta^{-1} & \zeta^2+\zeta^{-2} \end{pmatrix} \text{ where } \zeta = \zeta_5 = e^{2\pi i/5}.$$

Generators for the invariants:

$$y_1 = x_1^2 + \sum_{i=0}^{4} \left(\frac{x_1 + \zeta^i x_2 + \zeta^{-i} x_3}{\sqrt{5}} \right)^2$$

$$\begin{aligned} y_2 &= x_1^6 + \sum_{i=0}^{4} \left(\frac{x_1 + \zeta^i x_2 + \zeta^{-i} x_3}{\sqrt{5}} \right)^6 \\ y_3 &= x_1^{10} + \sum_{i=0}^{4} \left(\frac{x_1 + \zeta^i x_2 + \zeta^{-i} x_3}{\sqrt{5}} \right)^{10} \end{aligned}$$

and the Jacobian (of degree 15)

$$y_4 = \det \left(\frac{\partial y_i}{\partial x_j} \right)_{1 \le i,j \le 3}.$$

Relation: y_4^2 is some polynomial in y_1, y_2 and y_3.

Poincaré series $(1+t^{15})/(1-t^2)(1-t^6)(1-t^{10})$.

To obtain the reflection group described in entry 23 of Table 7.1, the scalar matrix with diagonal entries -1 should be adjoined, to extend the group to the full automorphism group $\mathbf{Z}/2 \times A_5$ of the icosahedron. The invariants of this group are generated by y_1, y_2 and y_3, which satisfy no relations.

(vii) $G \cong GL_3(\mathsf{F}_2)$ generated by the matrices

$$\begin{pmatrix} \zeta & 0 & 0 \\ 0 & \zeta^2 & 0 \\ 0 & 0 & \zeta^4 \end{pmatrix} \text{ and } \begin{pmatrix} \zeta^5 - \zeta^2 & \zeta^6 - \zeta & \zeta^3 - \zeta^4 \\ \zeta^6 - \zeta & \zeta^3 - \zeta^4 & \zeta^5 - \zeta^2 \\ \zeta^3 - \zeta^4 & \zeta^5 - \zeta^2 & \zeta^6 + \zeta \end{pmatrix} \text{ where } \zeta = \zeta_7 = e^{2\pi i/7}.$$

Generators of degrees 4, 6, 14 and 21 for the invariants:

$$\begin{aligned} y_1 &= x_1 x_2^3 + x_2 x_3^3 + x_3 x_1^3 \\ y_2 &= \det \left(\frac{\partial^2 y_1}{\partial x_i \partial x_j} \right)_{1 \le i,j \le 3} \\ y_3 &= \det \begin{pmatrix} \dfrac{\partial^2 y_1}{\partial x_i \partial x_j} & \dfrac{\partial y_2}{\partial x_i} \\ \dfrac{\partial y_2}{\partial x_j} & 0 \end{pmatrix} \\ y_4 &= \det \left(\frac{\partial y_i}{\partial x_j} \right)_{1 \le i,j \le 3}. \end{aligned}$$

Relation: y_4^2 is some polynomial in y_1, y_2 and y_3.

Poincaré series $(1+t^{21})/(1-t^4)(1-t^6)(1-t^{14})$.

To obtain the reflection group described in entry 24 of Table 7.1, the scalar matrix with diagonal entries -1 should be adjoined, to extend the group to $\mathbf{Z}/2 \times GL_3(\mathsf{F}_2)$. The invariants of this group are generated by y_1, y_2 and y_3, which satisfy no relations.

Appendix B

Examples over finite fields

(i) $G \cong \mathbb{Z}/3$ generated by the matrix $\begin{pmatrix} 0 & 1 \\ 1 & 1 \end{pmatrix}$ over $\mathbb{F}_2$.

Generators for the invariants:

$$\begin{aligned} y_1 &= x_1^2 + x_1x_2 + x_2^2 \\ y_2 &= x_1^3 + x_1^2x_2 + x_2^3 \\ y_3 &= x_1^3 + x_1x_2^2 + x_2^3. \end{aligned}$$

Relation: $y_1^3 = y_2^2 + y_2y_3 + y_3^2$.

Steenrod operations

	Sq^1	Sq^2
y_1	$y_2 + y_3$	y_1^2
y_2	y_1^2	y_1y_3
y_3	y_1^2	y_1y_2

Poincaré series $(1 + t^2 + t^4)/(1 - t^3)^2$.

The split extension of $(\mathbb{Z}/2)^2$ by this action of the cyclic group of order 3 is the alternating group A_4, and so the above ring (with the action of the Steenrod operations) is isomorphic to $H^*(A_4, \mathbb{F}_2)$. Furthermore, A_4 is a Sylow 2-normalizer in the alternating group A_5, and so this ring is also isomorphic to $H^*(A_5, \mathbb{F}_2)$ (see Section 5.1).

(ii) $G \cong \mathbb{Z}/4$ generated by the matrix $\begin{pmatrix} 0&1&0&0 \\ 0&0&1&0 \\ 0&0&0&1 \\ 1&0&0&0 \end{pmatrix}$ over $\mathbb{F}_2$.

Generators for the invariants:

$$\begin{aligned} y_1 &= x_1 + x_2 + x_3 + x_4 \\ y_2 &= x_1x_2 + x_2x_3 + x_3x_4 + x_4x_1 \\ y_3 &= x_1x_3 + x_2x_4 \\ y_4 &= x_1^2x_2 + x_2^2x_3 + x_3^2x_4 + x_4^2x_1 \\ y_5 &= x_1x_2x_3 + x_2x_3x_4 + x_3x_4x_1 + x_4x_1x_2 \\ y_6 &= x_1^2x_2x_3 + x_2^2x_3x_4 + x_3^2x_4x_1 + x_4^2x_1x_2 \\ y_7 &= x_1x_2x_3x_4 \end{aligned}$$

and a complicated list of relations.

Poincaré series

$$\frac{1 + 2t^3 + t^4}{(1-t)(1-t^2)^2(1-t^4)}.$$

This is an example of a ring of invariants which is not Cohen–Macaulay (Bertin [16]), and provided the first known example of a unique factorization domain which is not Cohen–Macaulay.

(iii) $G \cong A_n$, the alternating group of degree n, in its natural n dimensional representation. In characteristic zero and in odd characteristic, the invariants are generated by the elementary symmetric functions (see Section 1.1) together with the discriminant

$$\Delta = \prod_{i<j}(x_i - x_j)$$

whose square is a polynomial in the elementary symmetric functions. In characteristic two, however, Δ is already a polynomial in the elementary symmetric functions, because it is equal to

$$\Delta^+ = \prod_{i<j}(x_i + x_j).$$

The expression $(\Delta + \Delta^+)$, regarded as a polynomial with integer coefficients, is divisible by two, and so it makes sense to look at (the reduction modulo two of) the expression $(\Delta + \Delta^+)/2$. This, together with the elementary symmetric functions, generate the invariants in characteristic two (indeed, over any coefficient ring), and the square of this expression is a polynomial in the elementary symmetric functions.

Poincaré series $(1+t^{\binom{n}{2}})/\prod_{i=1}^{n}(1-t^i)$.

This is an example of the general fact that for a permutation representation of a group, the Poincaré series of the invariants is independent of the characteristic.

(iv) G is the group of upper triangular $n \times n$ matrices over F_q with ones on the diagonal, in its natural representation of dimension n over F_q.

Generators for the invariants:

$$
\begin{aligned}
y_1 &= x_1 \\
y_2 &= \prod_{a_1 \in \mathsf{F}_q} (a_1 x_1 + x_2) \\
y_3 &= \prod_{(a_1,a_2) \in (\mathsf{F}_q)^2} (a_1 x_1 + a_2 x_2 + x_3) \\
&\vdots \\
y_n &= \prod_{(a_1,\dots,a_{n-1}) \in (\mathsf{F}_q)^{n-1}} (a_1 x_1 + \cdots + a_{n-1} x_{n-1} + x_n).
\end{aligned}
$$

(No relations)

Poincaré series $1/\prod_{i=0}^{n-1}(1-t^{q^i})$.

(v) $G = GL_n(\mathbf{F}_q)$ in its natural representation of dimension n over F_q: see Section 8.1.

(vi) $G = SL_n(\mathbf{F}_q)$ in its natural representation of dimension n over F_q: see Section 8.2.

(vii) $G = Sp_{2n}(\mathbf{F}_q)$ in its natural representation of dimension $2n$ over F_q: see Section 8.3.

(viii) As a particular example of (v), we take $G = GL_3(\mathsf{F}_2)$ in its natural representation of dimension three over F_2.

Generators for the invariants:

$$
\begin{aligned}
c_2 &= x_1^4 + x_2^4 + x_3^4 + x_1^2x_2^2 + x_1^2x_3^2 + x_2^2x_3^2 + x_1^2x_2x_3 + x_1x_2^2x_3 + x_1x_2x_3^2 \\
c_1 &= x_1^4x_2x_3 + x_1x_2^4x_3 + x_1x_2x_3^4 + x_1^2x_2^2x_3^2 \\
&\quad + x_1^4x_2^2 + x_1^2x_2^4 + x_1^4x_3^2 + x_1^2x_3^4 + x_2^4x_3^2 + x_2^2x_3^4 \\
c_0 &= x_1x_2x_3(x_1+x_2)(x_1+x_3)(x_2+x_3)(x_1+x_2+x_3) \\
&= x_1^4x_2^2x_3 + x_1^4x_2x_3^2 + x_1^2x_2^4x_3 + x_1^2x_2x_3^4 + x_1x_2^4x_3^2 + x_1x_2^2x_3^4.
\end{aligned}
$$

(No relations)

Steenrod operations

	Sq^1	Sq^2	Sq^4
c_2	0	c_1	c_2^2
c_1	c_0	0	c_1c_2
c_0	0	0	c_0c_2

Poincaré series $1/(1-t^4)(1-t^6)(1-t^7)$.

(ix) G is the non-abelian Frobenius group of order 21, generated as a subgroup of $GL_3(\mathbb{F}_2)$ by the matrices $\begin{pmatrix} 0 & 1 & 1 \\ 1 & 0 & 0 \\ 1 & 0 & 1 \end{pmatrix}$ and $\begin{pmatrix} 0 & 1 & 0 \\ 0 & 0 & 1 \\ 1 & 0 & 0 \end{pmatrix}$.

Generators for the invariants: c_2, c_1, c_0 (see previous example) together with

$$\begin{aligned} y_1 &= x_1^3 + x_2^3 + x_3^3 + x_1x_2^2 + x_2x_3^2 + x_3x_1^2 + x_1x_2x_3 \\ y_2 &= x_1^5 + x_2^5 + x_3^5 + x_1x_2^4 + x_2x_3^4 + x_3x_1^4 + x_1^2x_2^2x_3 + x_1^2x_2x_3^2 + x_1x_2^2x_3^2. \end{aligned}$$

Relations:

$$\begin{aligned} y_1^4 &= c_2^3 + c_1^2 + y_1^2c_1 + y_2c_0 \\ y_2^2 &= c_1c_2 + y_1^2c_2 + y_1c_0. \end{aligned}$$

Steenrod operations

	Sq^1	Sq^2	Sq^4
y_1	c_2	y_2	0
y_2	y_1^2	c_0	$y_1c_1 + y_2c_2$

Poincaré series

$$\frac{(1+t^3)(1+t^5)(1+t^6)}{(1-t^4)(1-t^6)(1-t^7)}.$$

The split extension of $(\mathbb{Z}/2)^3$ by this action of the Frobenius group of order 21 is a Sylow 2-normalizer in the Janko sporadic group J_1, and so the above ring (with the action of the Steenrod operations) is isomorphic to $H^*(J_1, \mathbb{F}_2)$ (see Section 5.1).

(x) $G \cong SD_{16}$, the semidihedral group of order 16, generated by the matrices $\begin{pmatrix} 1 & 1 \\ -1 & 1 \end{pmatrix}$ and $\begin{pmatrix} 1 & 0 \\ 0 & -1 \end{pmatrix}$ over $\mathbb{F}_3$.

Generators for the invariants:

$$\begin{aligned} y_1 &= (x_1^2 + x_2^2)^2 \\ y_2 &= (x_1^2 + x_2^2)(x_1^4 + x_2^4) \\ y_3 &= (x_1^4 + x_2^4)^2. \end{aligned}$$

Relation: $y_2^2 = y_1 y_3$.

Poincaré series $(1+t^6)/(1-t^4)(1-t^8)$.

For the action of G on $\mathbb{F}_3[V] \otimes \Lambda(V^*)$ (cf. Section 5.1) the invariants are generated by y_1, y_2 and y_3 together with

$$\begin{aligned} u_1 &= (x_1^2 + x_2^2)d(x_1^2 + x_2^2) \\ u_2 &= (x_1^2 + x_2^2)d(x_1^4 + x_2^4) \\ u_3 &= (x_1^4 + x_2^4)d(x_1^2 + x_2^2) \\ u_4 &= (x_1^4 + x_2^4)d(x_1^4 + x_2^4) \\ v &= d(x_1^2 + x_2^2) \wedge d(x_1^4 + x_2^4). \end{aligned}$$

Relations $y_2^2 = y_1 y_3$, $u_i^2 = 0$, $u_i v = 0$ $(1 \le i \le 4)$, $v^2 = 0$ together with

$$\begin{aligned} u_1 u_2 &= y_1 v \\ u_1 u_3 &= 0 \\ u_1 u_4 &= y_2 v \\ u_2 u_3 &= y_2 v \\ u_2 u_4 &= 0 \\ u_3 u_4 &= y_3 v. \end{aligned}$$

Poincaré series

$$\frac{(1+t^3)(1+t^4)(1+t^7)(1+t^8) + (1-t^3)(1-t^4)(1-t^7)(1-t^8)}{2(1-t^8)(1-t^{16})}.$$

The split extension of $(\mathbb{Z}/3)^2$ by this action of SD_{16} is a Sylow 3-normalizer in the Mathieu sporadic groups M_{11} and M_{23}, and so the above ring is isomorphic to $H^*(M_{11}, \mathbb{F}_3)$ and also to $H^*(M_{23}, \mathbb{F}_3)$ (see Section 5.1).

(xi) G is a split extension of $(\mathbb{Z}/(p-1))^2$ (p odd) by $\mathbb{Z}/2$ (swapping the two copies of $\mathbb{Z}/(p-1)$), generated by the matrices $\begin{pmatrix} a & 0 \\ 0 & 1 \end{pmatrix}$ and $\begin{pmatrix} 0 & 1 \\ 1 & 0 \end{pmatrix}$ over $\mathbb{F}_p$, with a a primitive $(p-1)$st root of unity in $\mathbb{F}_p$.

Generators for the invariants:

$$\begin{aligned} y_1 &= x_1^{p-1} + x_2^{p-1} \\ y_2 &= x_1^{p-1} x_2^{p-1}. \end{aligned}$$

(No relations)

Poincaré series $1/(1-t^{p-1})(1-t^{2p-2})$.

According to Theorem 7.3.1 we have

$$(\mathsf{F}_p[V] \otimes \Lambda(V^*))^G = \mathsf{F}_p[y_1, y_2] \otimes \Lambda(dy_1, dy_2).$$

Poincaré series $(1-t^{2p-3})(1-t^{4p-5})/(1-t^{2p-2})(1-t^{4p-4})$.

The split extension of $(\mathbb{Z}/p)^2$ by this action of G is a Sylow p-normalizer in the symmetric group Σ_{2p}, and so the above polynomial tensor exterior algebra is isomorphic to $H^*(\Sigma_{2p}, \mathsf{F}_p)$ (see Section 5.1).

Bibliography

[1] A. Adem, J. Maginnis and R. J. Milgram. *Symmetric invariants and cohomology of groups.* Math. Ann. **287** (1990), 391–411.

[2] A. Adem and R. J. Milgram. *A_5-invariants, the cohomology of $L_3(4)$ and related extensions.* Proc. L.M.S. **66** (1993), 187–224.

[3] A. Adler. *Invariants of $PSL_2(\mathsf{F}_{11})$ acting on* $\mathbf{C}^5$. Commun. in Algebra **20** (1992), 2837–2862.

[4] G. Almkvist. *Invariants of $\mathbb{Z}/p\mathbb{Z}$ in characteristic p.* Proceedings of the 1982 Montecatini Invariant Theory Conference, Springer Lecture Notes in Mathematics **996**, Springer-Verlag, Berlin/New York, 1983.

[5] G. Almkvist and R. Fossum. *Decompositions of exterior and symmetric powers of indecomposable $\mathbb{Z}/p\mathbb{Z}$-modules in characteristic p and relations to invariants.* Séminaire d'Algèbre Paul Dubreil, Proceedings, Paris 1976–7, Springer Lecture Notes in Mathematics **641**. Springer-Verlag, Berlin/New York 1978.

[6] A.M.S. Proceedings of Symposia in Pure Mathematics XXVIII. *Mathematical developments arising from Hilbert problems.* A.M.S., 1976.

[7] M. F. Atiyah and I. G. Macdonald. *Introduction to commutative algebra.* Addison-Wesley, Reading, Mass. (1969).

[8] M. Auslander. *On the purity of the branch locus.* Amer. J. of Math. **84** (1962), 116–125.

[9] M. Auslander and D. Buchsbaum. *Unique factorization in regular local rings.* Proc. Nat. Acad. Sci. USA **45** (1959), 733–734.

[10] M. Auslander and D. Buchsbaum. *On ramification theory in Noetherian rings.* Amer. J. Math. **81** (1959), 749–765.

[11] M. Auslander and I. Reiten. *Grothendieck groups of algebras and orders.* J. Pure Appl. Algebra **39** (1986), 1–51.

[12] L. L. Avramov. *Pseudo-reflection group actions on local rings.* Nagoya Math. J. **88** (1982), 161–180.

[13] D. J. Benson and J. F. Carlson. *Projective resolutions and Poincaré duality complexes.* To appear in Trans. A.M.S.

[14] D. J. Benson and W. W. Crawley-Boevey. *A ramification formula for Poincaré series, and a hyperplane formula in modular invariant theory.* Preprint, 1993.

[15] M.-J. Bertin. *Anneau des invariants du groupe alterné, en caractéristique* 2. Bull. Sci. Math. **94** (1970), 65–72.

[16] M.-J. Bertin. *Anneaux d'invariants d'anneaux de polynômes, en caractéristique p.* C. R. Acad. Sci. Paris **277** (Série A) (1973), 691–694.

[17] N. Bourbaki. *Groupes et algèbres de Lie, Ch.* 4, 5 *et* 6. Masson, Paris 1981.

[18] C. Broto. *Àlgebres d'invariants i àlgebres sobre l'àlgebra de Steenrod.* Bull. Soc Cat. Cièn. VIII (1) (1986), 117–145.

[19] W. Burnside. *Theory of groups of finite order.* 2nd edition, C.U.P., 1911.

[20] H. E. A. Campbell, I. Hughes and R. D. Pollack. *Rings of invariants and p-Sylow subgroups.* Canadian Math. Bulletin **34** (1991), 42–47.

[21] D. Carlisle and P. Kropholler. *Modular invariants of finite symplectic groups.* Preprint.

[22] D. Carlisle and P. Kropholler. *Invariants of some classical finite groups over $GF(2)$.* Preprint.

[23] D. Carlisle and P. Kropholler. *Rational invariants of certain orthogonal and unitary groups.* Preprint.

[24] H. Cartan and S. Eilenberg. *Homological Algebra.* Princeton University Press, 1956.

[25] J. W. S. Cassels and A. Fröhlich. *Algebraic Number Theory.* Thompson Book Co. Inc., 1967.

[26] C. Chevalley. *Invariants of finite groups generated by reflections.* Amer. J. Math. **67** (1955), 778–782.

[27] A. Clark and J. Ewing. *The realization of polynomial algebras as cohomology rings.* Pacific J. of Math. **50** (1974), 425–434.

[28] H. S. M. Coxeter. *Discrete groups generated by reflections.* Ann. Math. **35** (1934), 588–621.

[29] M. Demazure. *Invariants symétriques entiers des groupes de Weyl et torsion.* Inventiones Math. **21** (1973), 287–301.

[30] W. Dicks and E. Formanek. *Poincaré series and a problem of S. Montgomery.* Linear and Multilinear Algebra **12** (1982), 21–30.

[31] L. E. Dickson. *A fundamental system of invariants of the general modular linear group with a solution of the form problem.* Trans. A.M.S. **12** (1911), 75–98.

[32] J. Dieudonné and J. B. Carrell. *Invariant theory, old and new.* Academic Press, New York 1971.

[33] J. Dixmier. *Sur les invariants du groupe symétrique dans certaines représentations.* J. Algebra **103** (1986), 108–192.

[34] J. Dixmier. *Sur les invariants du groupe symétrique dans certaines représentations, II.* In: Topics in invariant theory, Sém. d'Algèbre P. Dubreil, 1–34, Springer Lecture Notes in Mathematics **1478**, Springer-Verlag, Berlin/New York.

[35] A. Dress. *On finite groups generated by pseudoreflections.* J. Algebra **11** (1969), 1–5.

[36] G. Ellingsrud and T. Skjelbred. *Profondeur d'anneaux d'invariants en caracteristique p.* Compositio Math. **41** (1980), 233–244.

[37] M. M. Feldstein. *Invariants of the linear group modulo p^k.* Trans. A.M.S. **25** (1923), 223–238.

[38] N. L. Gordeev. *Finite linear groups whose algebra of invariants is a complete intersection.* Izv. Akad. Nauk SSSR Ser. Mat. **50** (1986), 343–392.

[39] P. Gordan. *Beweis, daß jede Covariante und Invariante einer binären Form eine ganze Function mit numerischen Coefficienten einer endlichen Anzahl solcher Formen ist.* Journal für die reine und angewandte Mathematik **69** (1868), 323–354.

[40] L. C. Grove and C. T. Benson. *Finite Reflection Groups.* Graduate Texts in Mathematics **99**, Springer-Verlag, Berlin/New York 1985.

[41] W. Haboush. *Reductive groups are geometrically reductive.* Ann. of Math. **102** (1975), 67–83.

[42] J. Herzog and E. Kunz, ed. *Der kanonische Modul eines Cohen–Macaulay-rings.* Springer Lecture Notes in Mathematics **238**, Springer-Verlag, Berlin/New York 1971.

[43] J. Herzog, E. Marcos and R. Waldi. *On the Grothendieck group of a quotient singularity defined by a finite abelian group.* J. Algebra **149** (1992), 122–138.

[44] J. Herzog and H. Sanders. *The Grothendieck group of invariant rings and of simple hypersurface singularities.* In: Singularities, Representation of Algebras, and Vector Bundles. Proceedings, Lambrecht 1985, 134–149. Springer Lecture Notes in Mathematics **1273**, Springer-Verlag, Berlin/New York 1987.

[45] D. Hilbert. *Ueber die Theorie der Algebraischen Formen.* Math. Ann. **36** (1890), 473–534.

[46] P. J. Hilton and U. Stammbach. *A course in homological algebra.* Graduate Texts in Mathematics **4**, Springer-Verlag, Berlin/New York 1971.

[47] M. Hochster and J. A. Eagon. *Cohen–Macaulay rings, invariant theory, and the generic perfection of determinantal loci.* Amer. J. Math. **93** (1971), 1020–1058.

[48] M. Hochster and J. L. Roberts. *Rings of invariants of reductive groups acting on regular rings are Cohen–Macaulay.* Advances in Math. **13** (1974), 115–175.

[49] W. C. Huffman and N. J. A. Sloane. *Most primitive groups have messy invariants.* Advances in Math. **32** (1979), 118–127.

[50] J. E. Humphreys. *Introduction to Lie algebras and representation theory.* Graduate Texts in Mathematics **9**, Springer-Verlag, Berlin/New York 1972.

[51] W. Hürlimann. *Sur le groupe de Brauer d'un anneau de polynômes en caracteristique p et la théorie des invariants.* Thesis, Zürich University, 1980.

[52] F. Ischebeck. *Eine Dualität zwischen den Funktoren Ext und Tor.* J. Algebra **11** (1969), 510–531.

[53] M.-C. Kang. *Picard groups of some rings of invariants.* J. Algebra **58** (1979), 455–461.

[54] M. Kervaire and T. Vust. *Fractions rationnelles invariantes par un groupe fini: quelques exemples.* In: Algebraische Transformationsgruppen und Invariantentheorie. Ed. H. Kraft, P. Slodowy and T. A. Springer. DMV Seminar **13**, Birkhäuser Verlag, Basel 1989.

[55] P. Landweber and R. Stong. *The depth of rings of invariants over finite fields.* Proc. New York Number Theory Seminar, 1984, Springer Lecture Notes in Mathematics **1240** (1987).

[56] I. G. Macdonald. *Symmetric functions and Hall polynomials.* Oxford Mathematical Monographs, Oxford University Press 1979.

[57] S. Mac Lane. *Homology.* Springer-Verlag, Berlin/New York 1974.

[58] H. Maschke. *The invariants of a group of* 2.168 *linear quaternary substitutions.* International Mathematical Congress 1893 (New York, Macmillan 1896), 175–186.

[59] H. Matsumura. *Commutative ring theory.* C.U.P., 1986.

[60] R. J. Milgram and S. B. Priddy. *Invariant theory and $H^*(GL_n(\mathbb{F}_p);\mathbb{F}_p)$.* J. Pure Appl. Algebra **44** (1987), 291–302.

[61] T. Molien. *Über die Invarianten der linearen Substitutionsgruppen.* Sitzungsber. König. Preuss. Akad. Wiss. (1897), 1152–1156.

[62] H. Morikawa. *On the invariants of finite nilpotent groups.* Osaka Math. J. **10** (1958), 53–56.

[63] H. Mui. *Modular invariant theory and cohomology algebras of symmetric groups.* J. Fac. Sci. Univ. Tokyo Sect. 1A Math. **34** (1987), 699–707.

[64] D. Mumford. *Geometric invariant theory.* Springer-Verlag, Berlin/New York, 1965.

[65] M. Nagata. *On the 14th problem of Hilbert.* Amer. Journal of Math. **81** (1959), 766–772.

[66] M. Nagata. *Local rings.* John Wiley & Sons, New York 1962.

[67] M. Nagata. *Invariants of a group in an affine ring.* J. Math. Kyoto Univ. **3** (1964), 369–377.

[68] H. Nakajima. *Invariants of reflection groups in positive characteristics.* Proc. Japan Acad. Ser. A **55** (1979), 219–221.

[69] H. Nakajima. *Invariants of finite groups generated by pseudoreflections in positive characteristic.* Tsukuba J. Math. **3** (1979), 109–122.

[70] H. Nakajima. *Invariants of finite abelian groups generated by transvections.* Tokyo J. Math. **3** (1980), 201–214.

[71] H. Nakajima. *On some invariant subrings of polynomial rings in positive characteristics.* Proc. 13th Symp. on Ring Theory, Okayama (1980), 91–107.

[72] H. Nakajima. *Modular representations of p-groups with regular rings of invariants.* Proc. Japan Acad. Ser. A **56** (1980), 469–473.

[73] H. Nakajima. *Modular representations of abelian groups with regular rings of invariants.* Nagoya Math. J. **86** (1982), 229–248.

[74] H. Nakajima. *Relative invariants of finite groups.* J. Algebra **79** (1982), 218–234.

[75] H. Nakajima. *Rings of invariants of finite groups which are hypersurfaces.* J. Algebra **80** (1983), 279–294.

[76] H. Nakajima. *Regular rings of invariants of unipotent groups.* J. Algebra **85** (1983), 253–286.

[77] H. Nakajima. *Rings of invariants of finite groups which are hypersurfaces, II.* Adv. in Math. **65** (1987), 39–64.

[78] E. Noether. *Der Endlichkeitssatz der Invarianten endlicher Gruppen.* Math. Ann. **77** (1916), 89–92.

[79] E. Noether. *Der Endlichkeitssatz der Invarianten endlicher linearer Gruppen der Characteristic p.* Nachr. Gött. Ges. Wissensch. (1926), 28–35.

[80] J. E. Olson. *A combinatorial problem on finite abelian groups, I.* J. Number Theory **1** (1969), 8–10.

[81] J. E. Olson. *A combinatorial problem on finite abelian groups, II.* J. Number Theory **1** (1969), 195–199.

[82] K. Pommerening. *Invariants of unipotent groups, a survey.* Invariant Theory, ed. Koh, Springer Lecture Notes in Mathematics **1278** (1987), 8–17.

[83] D. G. Quillen. *Projective modules over polynomial rings.* Inv. Math. **36** (1976), 167–171.

[84] P. Revoy. *Anneau des invariants du groupe alterné.* Bull. Sci. Math. **106** (1982), 427–431.

[85] D. J. Saltman. *Noether's problem over an algebraically closed field.* Invent. Math. **77** (1984), 71–84.

[86] P. Samuel. *Lectures on unique factorization domains.* Tata Institute Lecture Notes, Bombay 1964.

[87] B. J. Schmid. *Generating invariants of finite groups.* C. R. Acad. Sci. Paris **308**, Série I (1989), 1–6.

[88] B. J. Schmid. *Finite groups and invariant theory.* In: Topics in invariant theory, Sém. d'Algèbre P. Dubreil, 1–34, Springer Lecture Notes in Mathematics **1478**, Springer-Verlag, Berlin/New York.

[89] J.-P. Serre. *Faisceaux algébriques cohérents.* Ann. Math. **61** (1955), 197–278.

[90] J.-P. Serre. *Algèbre locale—multiplicités.* Springer Lecture Notes in Mathematics **11**, Springer-Verlag, Berlin/New York, 1965.

[91] J.-P. Serre. *Groupes finis d'automorphismes d'anneaux locaux réguliers.* Colloq. d'Alg. E.N.S. (1967).

[92] J.-P. Serre. *Local Fields.* Graduate Texts in Mathematics **67**, Springer-Verlag, Berlin/New York 1979.

[93] K. Shoda. *Über die Invarianten endlicher Gruppen linearer Substitutionen im Körper der Charakteristik p.* Jap. J. Math. **17** (1940), 109–115.

[94] B. Singh. *Invariants of finite groups acting on local unique factorization domains.* Journal of the Indian Math. Soc. **34** (1970), 31–38.

[95] G. C. Shephard and J. A. Todd. *Finite unitary reflection groups.* Canad. J. Math. **6** (1954), 274–304.

[96] L. Smith. *On the invariant theory of finite pseudo reflection groups.* Arch. Math. **44** (1985), 225–228.

[97] L. Smith and R. E. Stong. *On the invariant theory of finite groups: Orbit polynomials and splitting principals.* J. Algebra **110** (1987), 134–157.

[98] L. Solomon. *Invariants of finite reflection groups.* Nagoya J. of Math. **22** (1963), 57–64.

[99] L. Solomon. *Partition identities and invariants of finite groups.* J. Combinatory Theory (A) **23** (1977), 148–175.

[100] T. A. Springer. *Invariant theory.* Springer Lecture Notes in Mathematics **585**, Springer-Verlag, Berlin/New York 1977.

[101] R. P. Stanley. *Relative invariants of finite groups generated by pseudoreflections.* J. Algebra **49** (1977), 134–148.

[102] R. P. Stanley. *Invariants of finite groups and their applications to combinatorics.* Bull. A.M.S. **1** (3) (1979), 475–511.

[103] R. Steinberg. *Invariants of finite reflection groups.* Canad. J. Math. **12** (1960), 616–618.

[104] R. G. Swan. *Invariant rational functions and a problem of Steenrod.* Invent. Math. **7** (1969), 148–158.

[105] J. G. Thompson. *Invariants of finite groups.* J. Algebra **69** (1981), 143–145.

[106] J. S. Turner. *A fundamental system of invariants of a modular group of transformations.* Trans. A.M.S. **24** (1922), 129–134.

[107] K. Watanabe. *Certain invariant subrings are Gorenstein, I.* Osaka J. Math. **11** (1974), 1–8.

[108] K. Watanabe. *Certain invariant subrings are Gorenstein, II.* Osaka J. Math. **11** (1974), 379–388.

[109] K. Watanabe. *Invariant subrings which are complete intersections, I (invariant subrings of finite abelian groups).* Nagoya Math. J. **77** (1980), 89–98.

[110] H. Weyl. *Theorie der Darstellung kontinuierlicher halbeinfacher Gruppen durch lineare Transformationen.* Math. Zeit. **24** (1926), 377–395.

[111] H. Weyl. *The classical groups.* 2nd ed., Princeton Univ. Press, Princeton, 1953.

[112] C. Wilkerson. *A primer on the Dickson invariants.* Proc. of the Northwestern Homotopy Theory Conference, Contemp. Math. **19**, 421–434, A.M.S., 1983.

[113] O. Zariski and P. Samuel. *Commutative Algebra, Volume II.* Graduate Texts in Mathematics **29**, Springer-Verlag, Berlin/New York, 1975.

Index

Printed in the United States
By Bookmasters